Improved Modeling and Driving of Hydraulic Asymmetric Cylinders Systems

Wang Huankun 著
王焕昆

Beijing
Metallurgical Industry Press
2024

Metallurgical Industry Press
39 Songzhuyuan North Alley, Dongcheng District, Beijing 100009, China

图书在版编目(CIP)数据

改进的非对称液压缸系统建模与驱动=Improved Modeling and Driving of Hydraulic Asymmetric Cylinders Systems:英文/王焕昆著.—北京:冶金工业出版社,2024. 6(2024. 12 重印). —ISBN 978-7-5024-9895-5

Ⅰ. TH137. 51

中国国家版本馆 CIP 数据核字第 202468J3H4 号

Improved Modeling and Driving of Hydraulic Asymmetric Cylinders Systems

---

出版发行	冶金工业出版社	**电　话**	(010)64027926
地　址	北京市东城区嵩祝院北巷 39 号	**邮　编**	100009
网　址	www. mip1953. com	**电子信箱**	service@ mip1953. com

责任编辑　张熙莹　美术编辑　彭子赫　版式设计　郑小利
责任校对　王永欣　责任印制　窦　唯
北京建宏印刷有限公司印刷
2024 年 6 月第 1 版，2024 年 12 月第 2 次印刷
710mm×1000mm　1/16；9 印张；223 千字；132 页
定价 69. 00 元

投稿电话　(010)64027932　投稿信箱　tougao@cnmip. com. cn
营销中心电话　(010)64044283
冶金工业出版社天猫旗舰店　yjgycbs. tmall. com
(本书如有印装质量问题，本社营销中心负责退换)

Preface

This monograph is dedicated to enhancing the modeling and driving performance of asymmetric cylinder drive systems, with a focus on addressing challenges related to low-speed operation, velocity reversal, and friction phenomena. Our aim is to provide a comprehensive exploration of advanced modeling techniques, control methods, and innovative solutions tailored to the needs of hydraulic systems engineering. Drawing upon our expertise in hydraulic systems engineering, our intention in writing this book is to bridge the gap between theoretical understanding and practical implementation. We delve into the complexities of asymmetric cylinder systems, offering detailed analyses of their nonlinear behaviors and exploring strategies for improving their driving performance.

Key features of this book include:

(1) Detailed analysis of valve-controlled asymmetric cylinder systems, highlighting their nonlinear behaviors and performance under various operating conditions.

(2) Development of advanced simulation and analytical modeling techniques, facilitating faster and more accurate evaluation of system performance.

(3) Investigation and development of friction models tailored to hydraulic

systems, addressing the challenges of low-speed operation and velocity reversal.

(4) Introduction of a hybrid pump-controlled asymmetric cylinder system, aiming to optimize energy efficiency while maintaining control effectiveness.

(5) Extensive validation of theoretical models through experimental results, ensuring the practical applicability of proposed solutions.

The primary audience for this book includes:

(1) Hydraulic system engineers: professionals seeking to deepen their understanding of asymmetric cylinder drive systems and enhance their design and control capabilities.

(2) Hydraulic technology researchers: scholars and researchers dedicated to advancing the state-of-the-art in hydraulic system engineering.

(3) Control system professionals: engineers and researchers interested in applying advanced control techniques to hydraulic systems.

(4) Hydraulic system designers: engineers responsible for designing and optimizing hydraulic systems for various applications.

(5) Higher education scholars and students: university faculty and graduate students interested in the theoretical foundations and practical applications of hydraulic system engineering.

We acknowledge the collaborative efforts of all involved in the research, writing, and publication of this book. Their dedication and expertise have been instrumental in bringing this work to fruition.

In closing, we humbly express our gratitude to you, the reader, for choosing to engage with this book. We trust that it will provide valuable insights and practical knowledge to advance your understanding of asymmetric cylinder drive systems and inspire innovative solutions to real-world challenges.

Wang Huankun
January 20,2024

Contents

Chapter 1 From Historical Evolution to Disciplinary Contributions

This chapter contains overviews of relevant information about energy efficiency issues and solutions raised from hydraulic systems, and nonlinear phenomenon caused by the hydraulic system's internal structure design and external factors. Topics include analysis and comparison of various types of hydraulic designs for energy-saving purpose, their working principles, and the nonlinear behaviours explained with modeling and equations. Understanding and reviewing these pieces of this information are helpful to develop an energy efficiency hydraulic system circuit for this research and inhibit nonlinearities behaviours.

1.1 A wide variety of hydraulic systems

Hydraulic system applications meet the requirements of large forces, large torques and fast response in the industrial world. Though electro power transmission is increasingly popular, the great power-to-weight ratio of the hydraulic makes it still irreplaceable. Besides, the hydraulic system delivers power through fluid rather than rigid components like gears and shafts, so the system can be constructed compactly. The hydraulic covers a vast range of devices:

(1) Mining and transportation;

(2) Bridge, canal-barrage locks;

(3) Transport, road vehicles, rail, shipping, aircraft;

(4) Testing machines;

(5) Production line machines, injection modeling;

(6) Military vehicles, aerospace.

A typical application is an injection moulding machine, its hydraulic parts repeat extending and retracting duty cycles including mould closing, injecting molten material, holding mould and mould opening[1]. Fig. 1-1 depicts an injection moulding machine powered by hydraulic, the cylinder for screw-ram extends or retracts the screw to inject molten material, and the clamping screw operates the mould open or close.

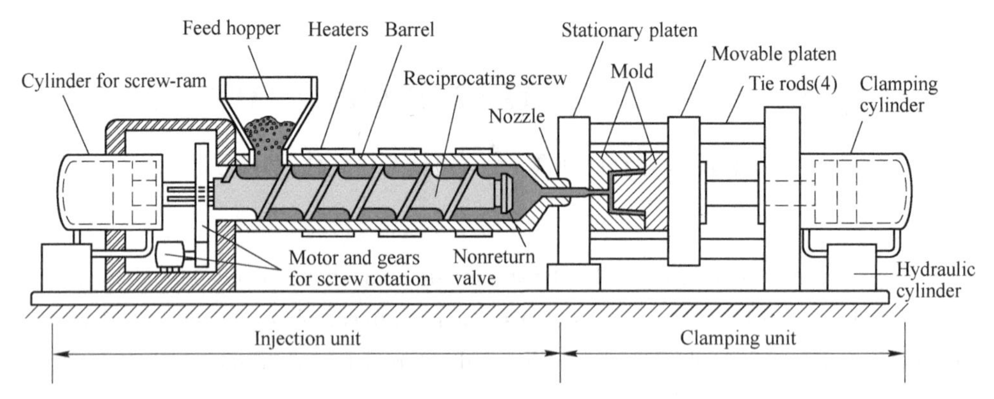

Fig. 1-1 Mould injection machine[2]

The hydraulic component in this type of machine is normally a valve-controlled cylinder system as shown in Fig. 1-2, in this type of injection moulding machine an asymmetric cylinder is used.

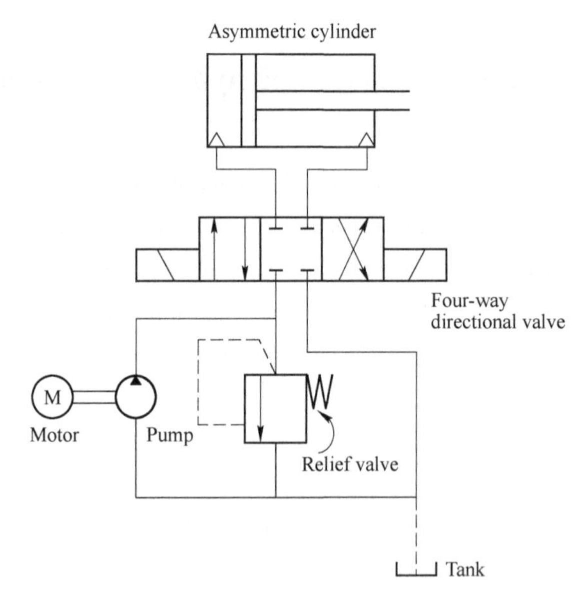

Fig. 1-2 Typical valve-controlled asymmetric cylinder system

In Fig. 1-2, in order to keep a constant supply pressure, the pump must keep working during cylinder operation, so that extra fluid will go back to the tank through the relief valve and the supply pressure is determined by the relief valve setting, such design leads to considerable energy losses[3]. Besides, the directional valve regulates the fluid flow by limiting its spool opening area, which causes throttle losses and a large amount of energy is transferred to heat.

An asymmetric cylinder is usually controlled by a symmetric ported valve, if the valve operates with a symmetric action, like a symmetric square wave, the asymmetric cylinder will respond with biased square wave motion, which causes oscillation behaviour when the cylinder changes its direction of motion[4]. Furthermore, the change of motion direction of the cylinder will affect the friction response[5], leading to more nonlinear phenomenon.

Viersma[6] proposed an asymmetric cylinder drive controlled by a symmetric four-way valve as in Fig. 1-3 to demonstrate its nonlinear behaviours.

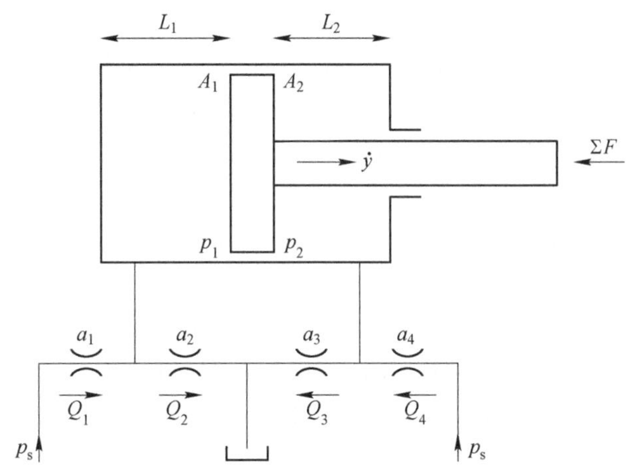

Fig. 1-3 Asymmetric cylinder controlled by a four-way valve scheme

A_1, A_2 —Piston areas; p_1, p_2 —Chamber pressures; p_s —Supply pressure; L_1, L_2 —The chamber length; $\dot{y}$ —Velocity; ΣF —Net force; a_1, a_2, a_3, a_4 —Valve opening areas; Q_1, Q_2, Q_3, Q_4 —The flows pass through their respective valve opening port; the area ratio $\frac{A_1}{A_2} = 2$

Viersma[6] depicts pressure jumps as in Fig. 1-4. Pressure jumps occur around $\dot{y} = 0$, causing "implosion" or "explosion" of oil owing to its compressibility, so that smooth operation around $\dot{y} = 0$ is impossible. Except for $A_1 = A_2$ (symmetric cylinder).

Most friction researches in hydraulics correlate the value of friction force with relative velocity and displacement between contacting surfaces only. For instance, Owen[7] utilises the LuGre friction model to describe friction in hydraulics. The cylinder piston seal deformation will occur when it is under pressure, which will also affect the friction

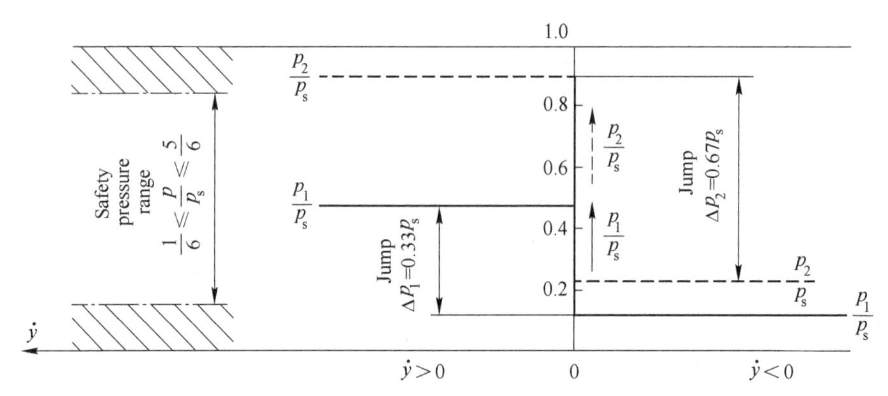

Fig. 1-4 Jump in cylinder pressure around $\dot{y} = 0$, due to asymmetric of the cylinder [6]

force behaviour[8].

Due to the difference between the piston area and rod side area of an asymmetric cylinder, the system is described by different models when it is in the extending or retracting state[9], which increases the hardship in performance analysis and system control.

As a summary for this section, hydraulic power transmission is still not replaceable due to its advantages, but with the growing requirements for energy efficiency, a conventional constant supply pressure valve-controlled hydraulic system can be redesigned to meet this demand. An energy-efficient design at low cost is one of the main targets of this research. Besides, the inevitable nonlinearities behaviour analysis of the asymmetric cylinder system and its friction under pressure are also a major concern.

1.2 Valves and pumps

According to the NFPA USA (National Fluid Power Association), the efficiency of fluid power systems is between the range of 4% to 60% and with an average efficiency of 22%. Hydraulic systems consist of many components which are responsible for the transfer of fluid within the systems and the transfer of power[10]. Even for a load-sensing hydraulic application, the energy losses in the valve can still be around 43% as in Fig. 1-5.

Energy efficiency is one of the concerns in industries in recent times as every company in the fluid power market wants to reduce its energy footprint to reduce energy costs and promote themselves as sustainable or so called "Green Manufacturers". Such systems are

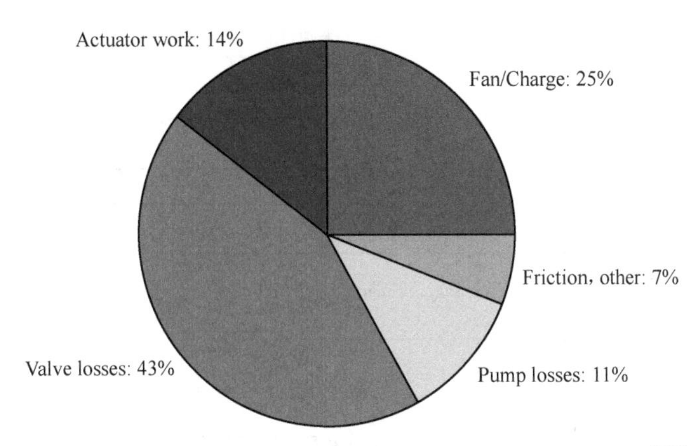

Fig. 1-5 Energy loss in mobile load sensing hydraulic application[10]

mainly separated into two types based on the design methodologies. One is the energy efficient valve-controlled system and the other is energy efficient pump-controlled system. While every solution has its pros and cons, benchmarks of energy efficiency for several design circuits under both concepts are carried out. Being aware of the characteristics of each type of design is helpful for future research.

1.2.1 Energy-efficient valve-controlled systems

The design of a valve-controlled system needs a valve to regulate the flow in the hydraulic system. Various methods are utilised to reduce the energy losses, for instance, load-sensing systems, energy recovery systems, digital hydraulic systems, etc.

The load-sensing hydraulic system is to reduce the energy losses by adjusting the pump outlet pressure to the highest working pressure required, a typical load-sensing hydraulic system is depicted in Fig. 1-6. This load-sensing system is composed of a swash-plate controlled axial-piston pump, a critical centred spool valve and an asymmetric cylinder.

This system measures the fluid pressure on the working side of the actuator and adjusts the swash-plate angle α to generate sufficient flow to maintain a pump discharge pressure p_P.

Levi[11] states that this load-sensing system provides less than 40% energy efficiency during retracting, but nearly 82% energy efficiency during cylinder extending. The efficiency difference is caused by the difference between the asymmetric cylinder piston areas, and utilising of the four-way valve to regulate the flow rate leads to throttle losses

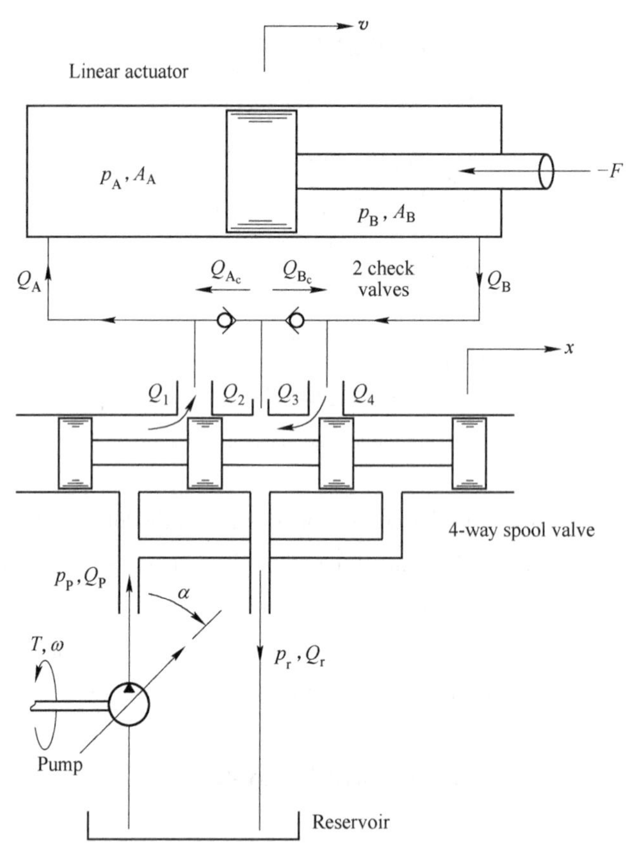

Fig. 1-6 Diagram of a typical load-sensing hydraulic system[11]

v — The cylinder velocity; p_A —The piston chamber pressure; p_B —The rod side chamber pressure;

A_A —The piston area; A_B —The rod side piston area;

Q_A —The flow delivered into the piston side chamber; Q_B —The flow delivered into the rod side chamber;

Q_{A_c}, Q_{B_c} —The compensate flow when the cylinder is overrunning;

Q_1, Q_2, Q_3, Q_4 —The flow through the corresponding valve port respectively;

p_P —The pump outlet pressure; Q_P —The pump outlet flow; p_r —The pressure in reservoir;

Q_r —The flow returning to the reservoir; T —The motor torque applied on the pump;

ω — The angular speed and α is the angle of the pump swash plate

in the system.

Wang[12] proposed a load-sensing system combined with energy recovery function as depicted in Fig. 1-7, the load-sensing pilot line adopts the highest working pressure of both actuators and adjusts the pump outlet pressure to the required value. In the working line with energy recovery, when the operation condition is overrunning or the inlet pressure is not the highest, the hydraulic motor and its generator convert the cylinder outlet energy into electrical form. The recovered energy can be restored in battery or supercapacitor or directly supplied to the electric required devices.

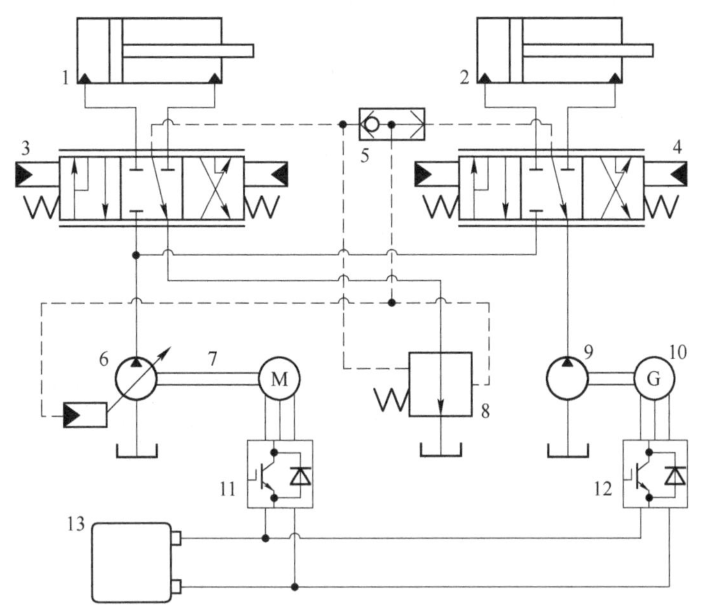

Fig. 1-7 Load-sensing & energy recover hydraulic system[12]

1—Cylinder (without energy recovery); 2—Cylinder (with energyrecovery); 3, 4—Control valve; 5—Shuttle valve; 6—Variabledisplacement pump; 7—Electric motor; 8—Compensator; 9—Hydraulicmotor; 10—Generator; 11, 12—Converter; 13—Super capacitor/battery

At last, the stored power can be recovered to supply devices that require electricity. The energy recovery efficiency is tested with a boom cylindersystem[13], which is described in Fig. 1-8. Its variable displacement pump is driven by a hybrid power engine, consisting of a diesel engine and an electrical machine, the pressure relief valve is used for limiting the supply pressure.

When the boom cylinder retracts, the gravity will drive the oil in the cylinder chamber to flow through the hydraulic motor, which actuates the coupling generator to produce electricity and store the power in the supercapacitor. When the boom cylinder extends, the stored power in the supercapacitor drives the cylinder motor (acting as a pump) to assist the oil flow back to the tank. Therefore, the power required by the active pump (next to the motor) and the fuel consumption of the system are reduced.

This energy recovery hydraulic device is tested under different conditions, the overall energy recovery ratio is between 26% and 33%[13]. Though this system is able to recover

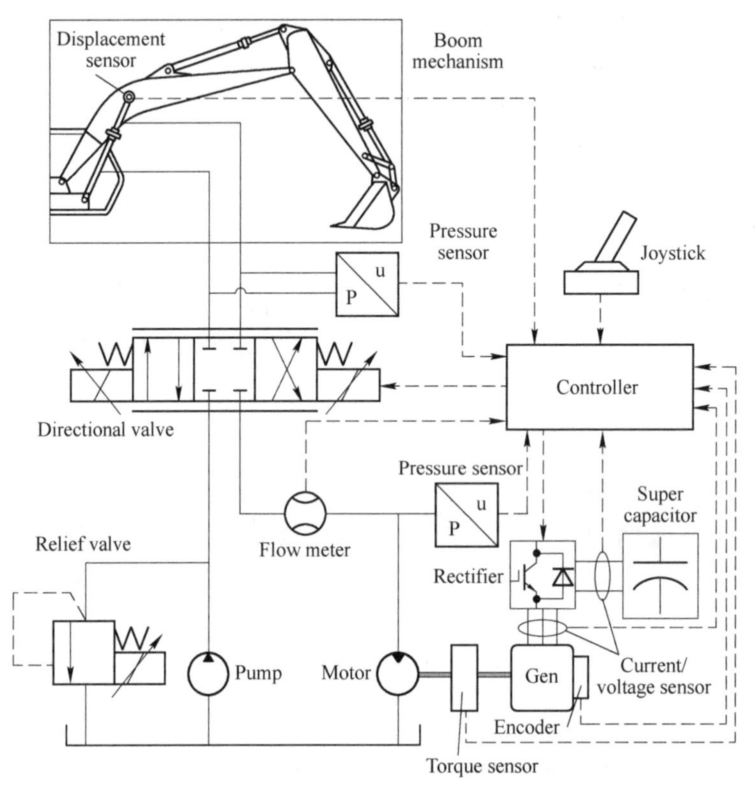

Fig. 1-8 Test bench of boom cylinder circuit with energy recovery[13]

energy at a certain ratio, the supply pressure is still fixed, which means there is extra fluid flow to the tank to maintain the supply pressure. The relief valve setting can be set to a required minimum value to reduce the energy losses, but it is still inevitable.

Based on the same energy recovery design strategy, an energy regeneration system is put forward by reference [14] as depicted in Fig. 1-9. The accumulator connected to the pump is mainly used for reducing the pressure oscillations from the pressure source, and the other one connected to the cylinder chamber is applied for energy recovery. The "variable damping" in Fig. 1-9 can be considered as a servo needle valve, when the cylinder is placed at left end stroke, the needle valve is almost closed, while in the right end stroke, it is opened at maximum.

The energy restoring and recovering processes are clear and simple. When the cylinder retracts to its left end, the hydraulic oil is compressed into the accumulator, and when the cylinder extends to the right end, the accumulator releases the pressed fluid into the chamber. Man[14] stated that the system with accumulators could reduce energy

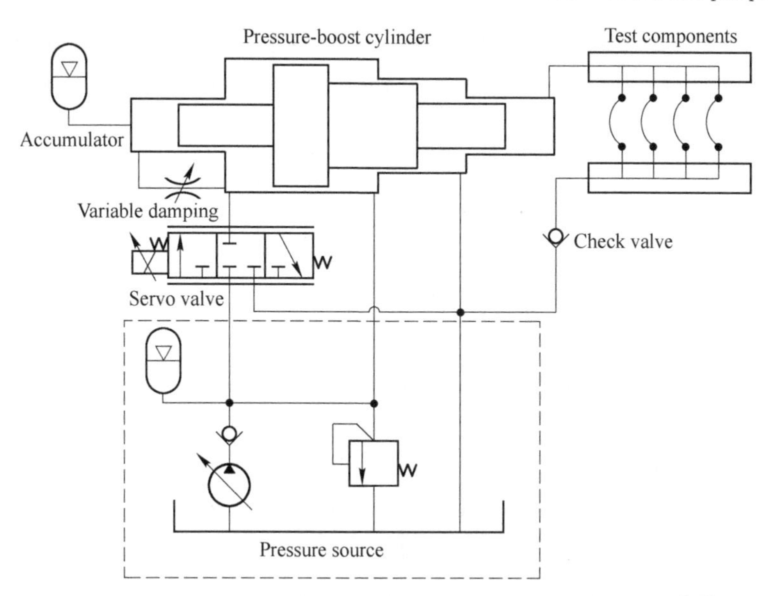

Fig. 1-9 Energy regeneration system for hydraulic impulse testing [14]

consumption by 15% compared with the system without an accumulator.

A Digital hydraulic valve system utilises a number of on/off valves for its operation. The on/off valves are part of the digital flow control unit as depicted in Fig. 1-10. The on/off valves may have equal flow capacity and the unit flow is controlled by pulse number modulation (PNM)[15].

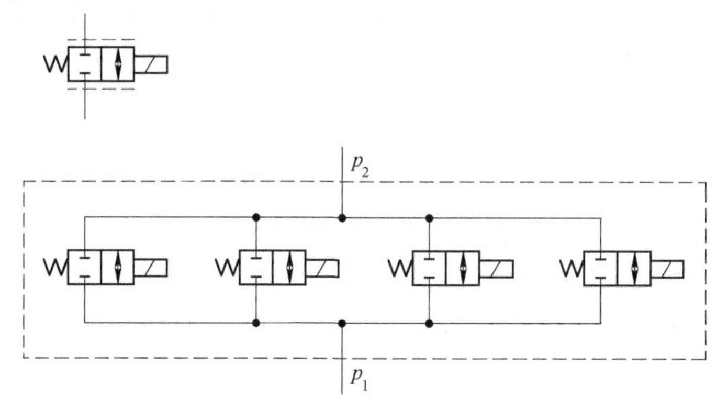

Fig. 1-10 Diagram of digital flow control unit (DFCU) and its drawing symbol [15]

There are usually a few (four or more) digital flow control units (DFCUs) in the hydraulic system to perform independent metering control. Fig. 1-11 presents a digital valve system consisting of four DFCUs. From the energy perspective, Huova[15] compares the energy losses of a digital valve system with a load-sensing four-way proportional valve system, the energy losses in the digital valve system are reduced by 29%-67% compared to a load-sensing proportional valve system in different load conditions.

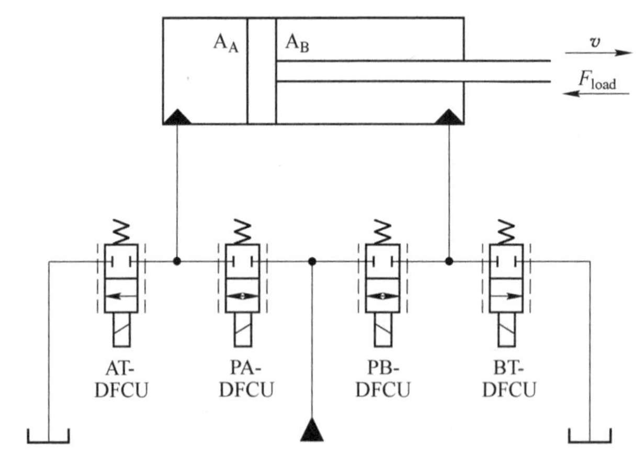

Fig. 1-11 Four-DFCU controlled digital valve system [15]

However, the control mechanism of the digital valve system is far more complicated, with multiple on/off valves existing in one single DFCU. Therefore, the states of the valves must be taken into consideration in the controlling method.

With these examples, various types of efficient valve-controlled systems are reviewed:

(1) The load-sensing system is to reduce the energy consumption from the input aspect, but throttle losses are inevitable.

(2) The energy recovery system aims at recycling energy and reusing it to power the system. Its reduction of energy consumption is around 15% to 30%. Applications can be found in references [16]-[18].

(3) Digital hydraulic systems have better energy efficiency compared to a conventional load-sensing system, due to its independent metering function, but its control method is far more complicated.

Throttle losses in these applications can still be found, which is inevitable in any valve-controlled hydraulic system. The next section discusses the pump-controlled system applications that eliminate throttle losses in hydraulics.

1.2.2 Pump-controlled hydraulic systems

Another type of energy-saving hydraulic design is to change its structure design to avoid throttling losses and reduce the fluid required, the pump directly controlled hydraulic systems meet these requirements. They can be summarized into two types of system[19], an open circuit and a closed circuit as depicted in Fig. 1-12. These are basic simplified circuits and some components are missing. Though the open circuit is still applied with a control valve, it usually operates from one end stroke to the other end, which indicates that the valve will always stay at its maximum opening state during operation[20]. The closed circuit does not even have a valve, the pump directly delivers fluid oil into the cylinder chamber, the pump must be a bidirectional type to be able to deliver fluid into different chambers[21].

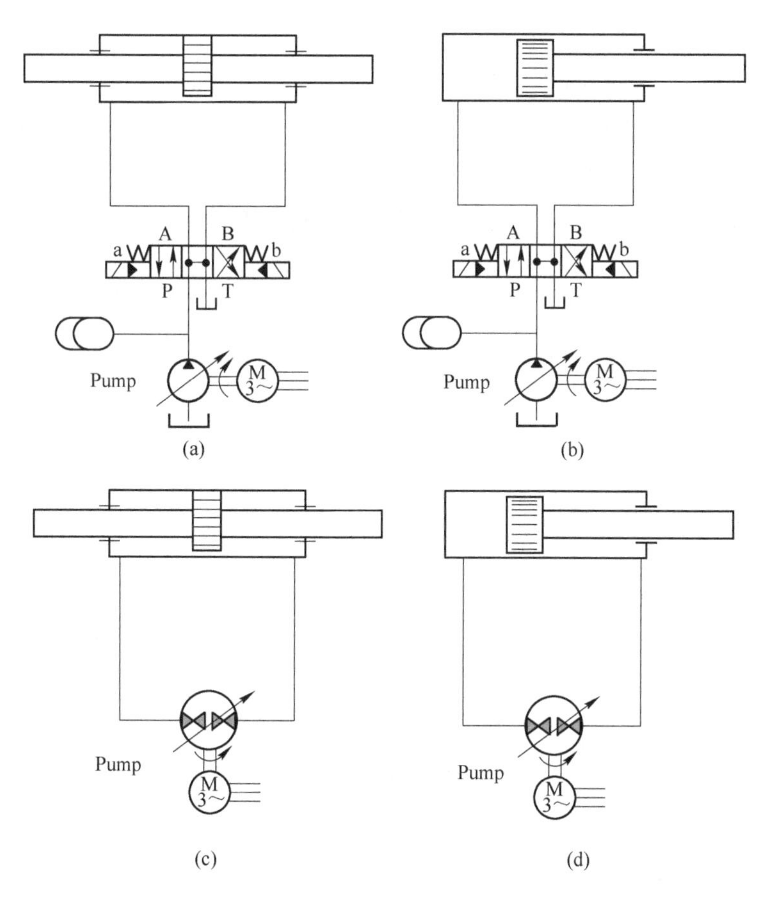

Fig. 1-12 Simplified basic schemes of open circuit (a and b) and closed circuit (c and d)[19]

Quan[19] reviewed and summarised that the pump-controlled hydraulic system's energy efficiency can be improved by more than 40% on average compared to a conventional valve-controlled system. However, as the circuits in Fig. 1-12 are simplified basic circuits, some problems can be found. The accumulators in the open circuit in Fig. 1-12 (a) and (b) intend to provide some charging pressure during operation, but the accumulators will affect the stability performance of the open circuit and the accumulators are not essential. For the closed circuit in Fig. 1-12 (d), imbalance flow problem is obvious, when the system is extending or retracting, cavitation will happen in one of the cylinder chambers and leads to a stability problem. Normally, check valves that connected to an oil tank will be added to this type of closed circuit to compensate for the imbalanced flow rate. Based on these basic circuits, various designs are put forward by researchers and they are reviewed in the rest part of this section.

An investigation of a unidirectional pump-controlled asymmetric cylinder position control system is carried out[22]. Fig. 1-13 shows the schematic circuit of the system. The system is based on the open circuit design using a control valve to switch the flow direction, the return line fluid is delivered back to the tank.

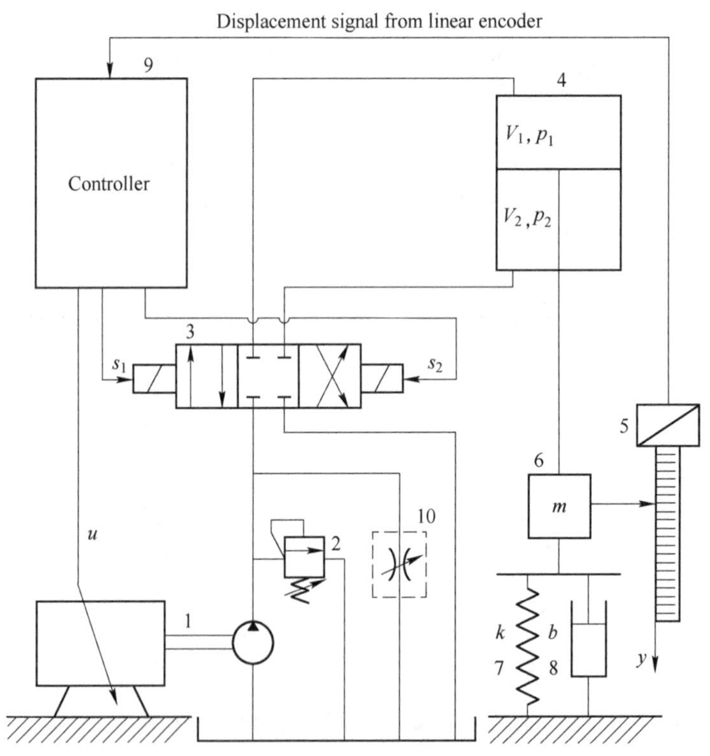

Fig. 1-13 Schematic circuit of unidirectional pump-controlled asymmetric cylinder system[22]

1—Pump; 2—Safty valve; 3—Direction control valve; 4—Asymmetric cylinder; 5—Displacement sensor; 6—Mass block; 7—Spring; 8—Damper; 9—Motion controller; 10—Needle valve; m—Mass; k—Spring coefficient; b—Damping; y—Displacement

The directional valve in this circuit only operates at left end stroke or right end stroke to maximize the opening area during operation to reduce the throttle losses. The cylinder velocity, position and chamber pressure are controlled by the AC servo motor only. However, in Tao's research[22], the controller only covers the cylinder extension scenario and its model predictive controller is constructed based on the mathematic model of the system when the cylinder extends. This kind of open circuit pump-controlled strategy is utilised in an injection moulding machine to achieve the energy-saving purpose[23] and it saves around 30% of energy compared to a conventional system. In a mobile pump-controlled hydraulic system[20], such a design reduces around 14% of energy compared to a load-sensing design.

For the closed circuit, Rahmfeld proposed a variable displacement pump-controlled asymmetric cylinder system[24] as depicted in Fig. 1-14, its pilot check valves are implemented to balance the flow differential during operation. The relief valve that is connected to the unidirectional pump is used to maintain the charge pressure. When any cylinder chamber pressure is lower than some certain value, compensation can be achieved through the check valve.

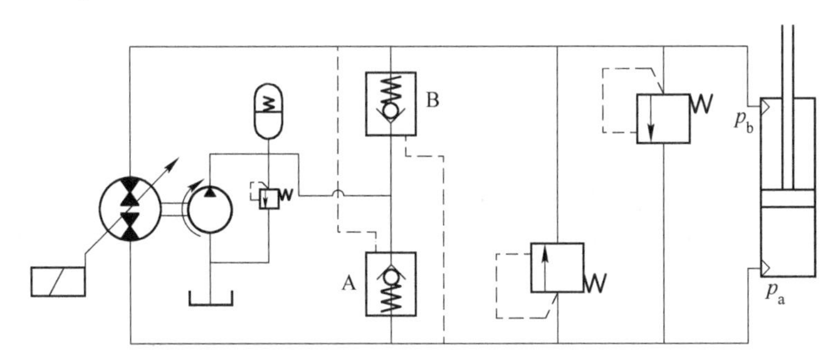

Fig. 1-14 Variable pump-controlled asymmetric cylinder system [24]

When the cylinder is extending, the chamber pressure p_a increases but chamber pressure p_b decreases much faster due to the unbalanced flow rate, during which the check valve B is opened and the unbalanced flow is compensated. When the cylinder is retracting, chamber pressure p_a increases rapidly due to the unbalanced flow, and pressure p_b must be larger to balance the force, so that the check valve A opens and extra flow from the piston chamber can be released. The latter causes p_a to decease and thus check valve A to close. The process of p_a increasing and decreasing, causing the check valve A to open and close, will be repeated and oscillations will occur in the system.

An improved design is carried out[25], which utilises two servo flow control valves to stabilise the system performances as depicted in Fig. 1-15. When oscillation occurs, the servo valves 4a and 4b will open to release the extra pressurised fluid into the tank. Wang [25] stated that his design encountered a critical mass problem.

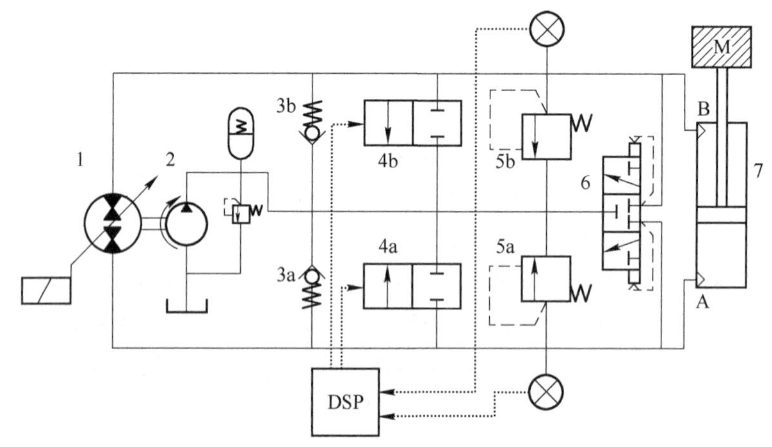

Fig. 1-15 Improved closed circuit of pump-controlled asymmetric cylinder system[25]

1—A displacement controlled pump; 2—Charge pump;

3a, 3b—A pair of check valves; 4a, 4b—A pair of flow control valves;

5a, 5b—A pair of relief valves; 6—A three-position three-way shuttle valve;

7—A single rod cylinder

If the load added to the cylinder is far from a critical value, the system performances are satisfied, but when the load is close to the critical value and the controller is off, pressure oscillations during operation become severe as described in Fig. 1-16.

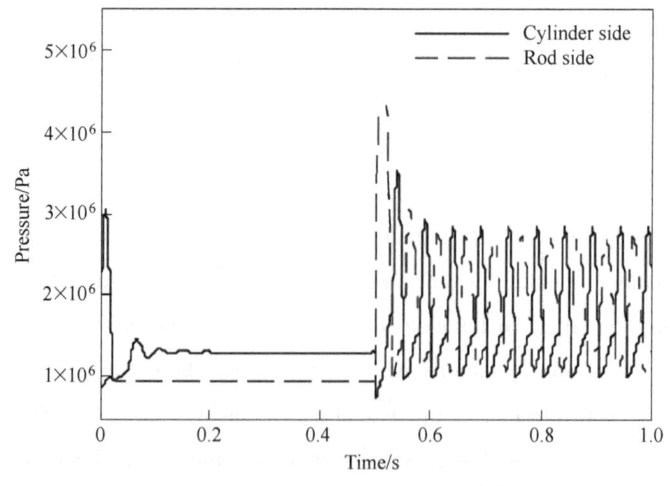

Fig. 1-16 Pressure oscillations [25]

In another paper[26], it claimed the pressure oscillation occurs when both cylinder chamber pressures are close to the charge pressure. It can be noticed that the severe pressure oscillations only happen when the cylinder is under the retracting state, which is caused by the check valve to open and close under this situation. Similarly, Williamson[27] observed oscillations in his closed circuit pump-controlled hydraulic system when the cylinder is rapidly lowering small loads.

A closed circuit pump-controlled system with an accumulator is simulated by[28] as in Fig. 1-17, its system energy efficiency performance varied from 32% to 66% depending on the pump or motor displacement, but its open loop test shows oscillations during operation.

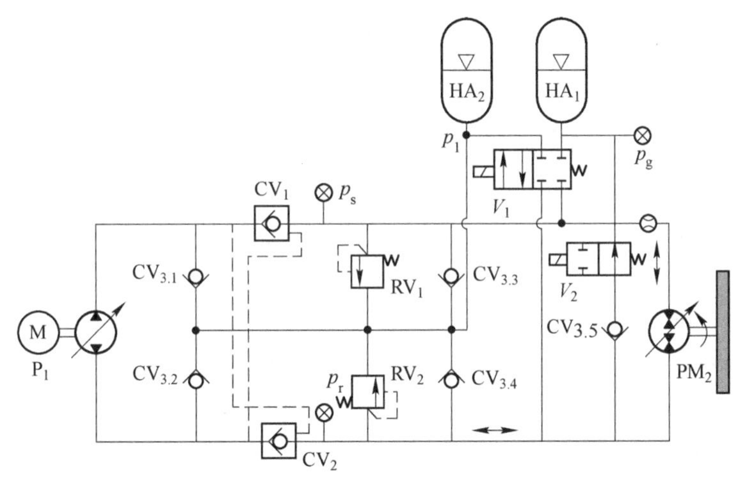

Fig. 1-17 A pump-controlled hydraulic system with energy recovery [28]

The oscillations are caused by the accumulator in the return line to charge and discharge in a step input test as in Fig. 1-18.

As throttle losses can be eliminated as there is no valve to regulate the flow rate and a constant supply pressure is not essential in a pump-controlled hydraulic system, the pump-controlled system usually has much better energy efficiency compared to a valve-controlled system.

Without a servo proportional valve, the flow rate and pressure control of the pump-controlled system are achieved by adjusting the pump speed, accumulator and relief valve settings. As these components do not dynamically fast respond like a servo valve, a pump-controlled system is likely to perform oscillations under some certain conditions. These factors lead to the requirement of a more complicated controller to eliminate oscillations.

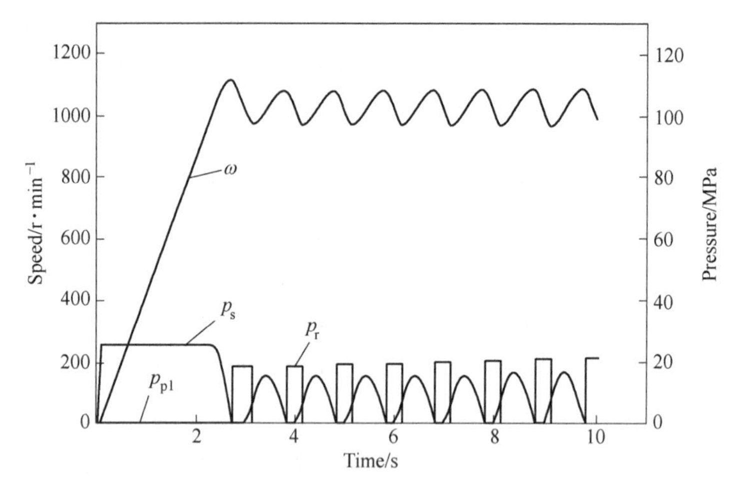

Fig. 1-18 Speed and pressure of the pump-controlled system[28]

As a summary of the energy efficiency solutions for hydraulic systems, they can be concluded into two major types, one is a valve-controlled system and the other is a pump-controlled system. Many pieces of research are not exactly the same as the examples in sections 1. 2. 1 and 1. 2. 2, but there is no substantial difference, Li[29] combined a hydraulic accumulator with an electric regeneration unit to achieve a better energy recovery solution. Zhao[30] proposed a hydraulic hybrid excavator system with an energy recovery accumulator added to a pump-controlled system, which offered great energy efficiency.

Compared between these designs and examples, the valve-controlled solutions can offer better controllability than pump-controlled designs, but the latter has greater energy efficiency. In pump-controlled designs, they can be sorted into the open circuit or closed circuit[19], the open circuit is more stable compared to the closed one. When the asymmetric cylinder is the actuator, the closed circuit must take flow compensation into consideration[31], which can cause some unexpected behaviours, like pressure oscillation. However, an open circuit is often equipped with a servovalve, which is cost-unfriendly. Hence, an ideal hydraulic actuator system should be able to offer good energy efficiency, low cost and good stability, which is one of the targets of this research.

Simulation of hydraulics helps to understand its system performance, nonlinear behaviours and working principles. The energy-efficient valve-controlled system is basically a conventional servovalve-controlled hydraulic system[32], but the pump-controlled system is not equipped with a constant supply pressure, so that their modeling equations and methods will be different. Two modeling examples for both types are reviewed, their equations and methods are useful for future simulation of this research.

1.2.3 Modeling of a pump-controlled hydraulic system

The last section reviewed a simulation strategy of a constant supply pressure servovalve controlled hydraulic system. A "Component Linking" simulation method performs well for that setup, however, the pump-controlled system does not operate with constant supply pressure. When operating at a constant pump speed, the pump will provide a constant flow rate to the system. Thus, the modeling of this type of hydraulic system is worth reviewing and evaluating.

1.2.3.1 A state space model for a pump-controlled asymmetric cylinder system

A unidirectional proportional pump-controlled asymmetric cylinder system test rig is proposed in reference [33], which is depicted in Fig. 1-19. The system consists of an AC servo motor, unidirectional variable speed proportional pump, directional control valve, asymmetric cylinder, oil tank, sensors, mass block, springs and damper.

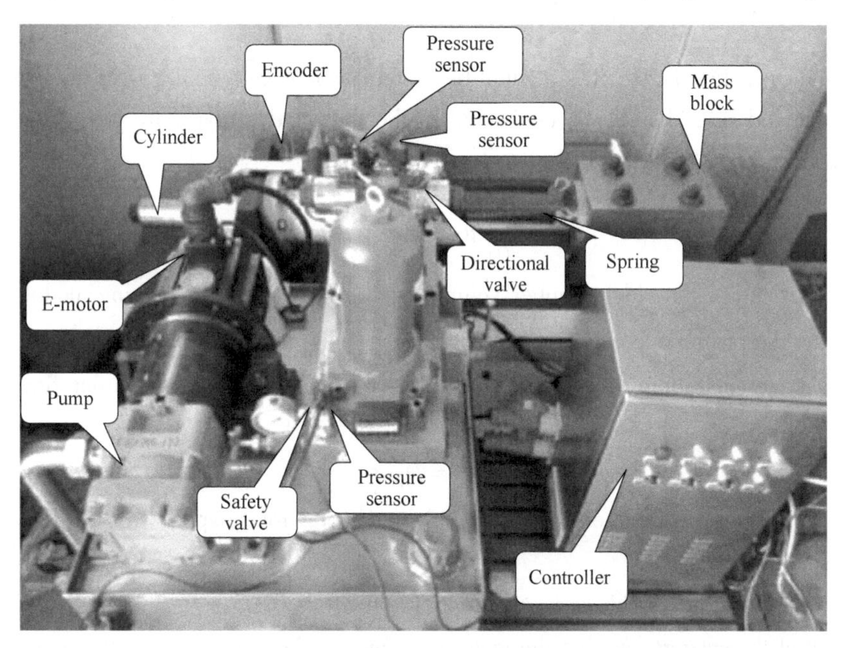

Fig. 1-19 A pump-controlled asymmetric cylinder test rig [33]

The circuit of this system is depicted in Fig. 1-19, with some assumptions proposed by reference [33], and the circuit can be simplified as in Fig. 1-20. Assuming the oil temperature and bulking modulus are constant, the pump rotary speed is proportional to the input signal, pressure drops of the directional valve are small and negligible, pump leakage is neglected, safety valve is closed during operation, the cylinder chamber's volume introduced by piston movement is small and negligible and the initial values of system states are zero.

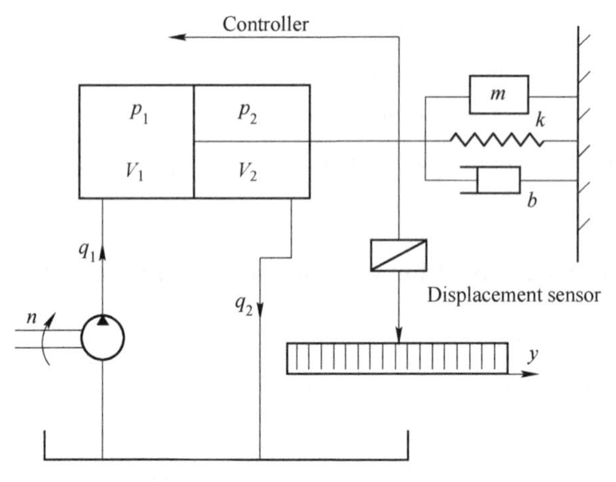

Fig. 1-20 Simplified circuit for pump-controlled asymmetric cylinder[33]

As shown in Fig. 1-20, it is a simplified circuit when the cylinder is in the extending state, the rod side chamber is connected to the tank and pressure $p_2 = 0$. Hence, the flow balance equation can be written as:

$$q_1 = \frac{V_1}{B_k}\dot{p}_1 + A_1\dot{y} + K_{ci}p_1 \tag{1-1}$$

where V_1 is the piston side chamber volume, m^3 ; B_k is the fluid bulking modulus, N/m^2; p_1 is the piston side chamber pressure, N/m^2; A_1 is the piston side area; K_{ci} is the inner leakage coefficient, $m^5/(N \cdot s)$; q_1 is equal to the pump's output flow Q_P, m^3/s , which is:

$$q_1 = D_P\omega \tag{1-2}$$

where D_P is the ratio between the pump flow and motor rotary speed, m^3/rad; ω is the pump speed, rad/s.

The time constant of the servo motor in Fig. 1-19 is 0.05 s [33], which is far less than that of the system, so that the motor dynamic can be ignored, and the rotary speed ω is:

$$\omega = K_{\omega} u \tag{1-3}$$

where u is the command signal, V; K_{ω} is the coefficient of the command signal to motor speed, rad/(s · V).

Considering the force balance on the load, the following equation is obtained:

$$p_1 A_1 = m\ddot{y} + b\dot{y} + ky \tag{1-4}$$

where m is the mass of the load, kg; b is the coefficient of damping, N · s/m; k is the stiffness of the spring, N/m; y is the displacement, m.

Define the state variables $x_1 = p_1$, $x_2 = p_2$, $x_3 = \dot{y}$ and the vector $X = [x_1 \quad x_2 \quad x_3]^{\mathrm{T}}$, $U = u$ is the input variable and $Y = y$ is the output variable. The state space model is given by:

$$\begin{cases} \dot{x} = AX + BU \\ Y = CX \end{cases} \tag{1-5}$$

where A, B and C is the coefficient matrix of the system, they are revealed as:

$$\begin{cases} A = \begin{bmatrix} -\dfrac{B_k}{V_1}K_{\mathrm{ci}} & 0 & -\dfrac{B_k}{V_1}A_1 \\ 0 & 0 & 1 \\ \dfrac{A_1}{m} & -\dfrac{k}{m} & -\dfrac{b}{m} \end{bmatrix} \\ B = \begin{bmatrix} \dfrac{B_k}{V_1}D_{\mathrm{P}}K_{\omega} \\ 0 \\ 0 \end{bmatrix} \\ C = [0 \quad 1 \quad 0] \end{cases} \tag{1-6}$$

The simulation model proposed in reference [33] is implemented only under cylinder extending state, the simulation model in the cylinder retracting state is not analysed and constructed.

For sure the state space model will be different due to the force balance equation changes when cylinder in the retracting state, this difference will lead to system nonlinear behaviours and also the nonlinearities for its control[34]. In a pump-controlled steer-by-wire system, two separate models are utilised to simulate this pump-controlled asymmetric cylinder system[35]. Similar strategies can be found in references [36] - [38].

1.2.3.2 A summary of the state space model

In the last section, a state space model for a pump-controlled asymmetric cylinder hydraulic system is carried out[33]. The cylinder chamber V_1 volume is set to be constant in coefficient matrix A. However, the chamber volume V_1 does not possibly remain constant during cylinder operation. The reason for using the constant coefficient matrix is to build a controller in an easier way, and the difference between the model and the real test rig can be compensated by the controller. However, such a model is not able to reflect all the nonlinearities in a pump-controlled asymmetric cylinder system, and the model only describes the system in the extending state. The flow balance equations of the pump-controlled asymmetric cylinder system are different in either extending or retracting state, which indicates the system parameters under both states are different as well. These points are worth to be noticed in future research.

1.3 Mysterious friction

Friction is inevitable in a fluid power system, which includes sliding components, especially for a hydraulic actuator. The majority of friction existing in a hydraulic system is produced by the cylinder seals and the relative motion of the cylinder piston, some undesired influences may affect the performance of the system. Thus, friction behaviours must be taken into consideration.

1.3.1 Basic friction behaviours

The friction force is generated from the relative motion between two contacting surfaces, so that most of the characteristics of the friction can be depicted as a function of the velocity. Their basic behaviours are depicted in Fig. 1-21[39].

Fig. 1-21 (a) describes Coulomb friction, which is a constant value due to the degree of roughness of the contacting surfaces. It is independent of the velocity and regarded as a simple model for dry friction.

Fig. 1-21 (b) illustrates the viscous friction which is proportional to velocity, and expressed as a function of coefficient B multiplied with the velocity as in the graph.

In Fig. 1-21 (c), a friction phenomenon called Stribeck effect is depicted, which introduces friction force at low velocity. It is decreasing exponentially from the difference between the stiction force and the Coulomb force to zero[40].

Fig. 1-21 (d) combines all of the three friction behaviours in the graph (a), (b) and (c), revealing the basic friction behaviours under different velocity ranges. A

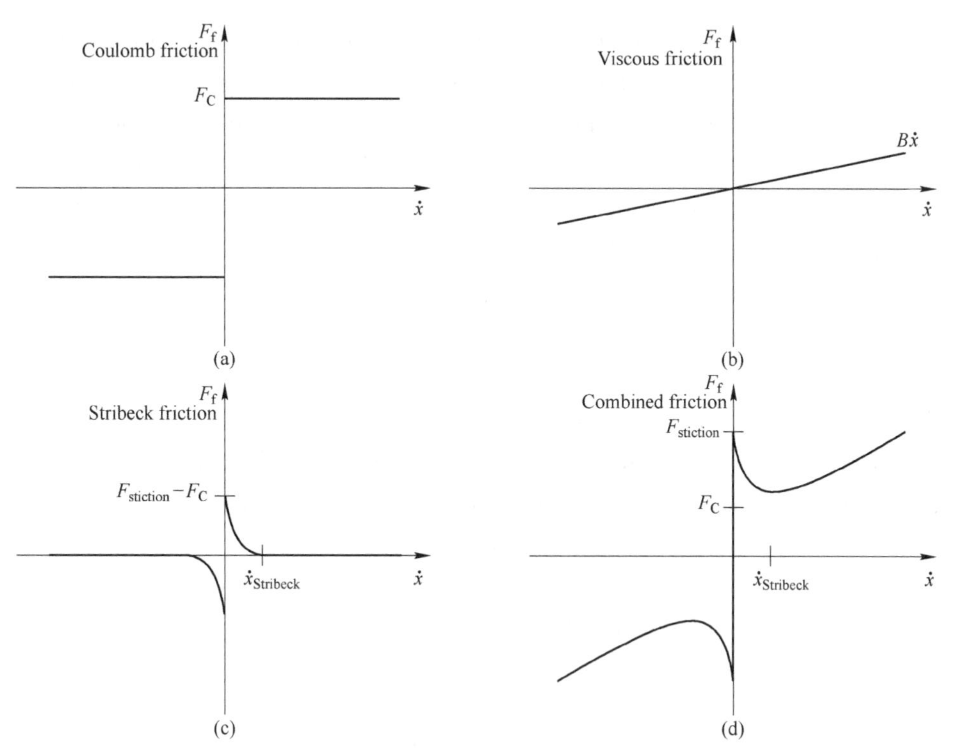

Fig. 1-21 General friction characteristics and the combined friction[39]

detailed explanation of the Stribeck regime is carried out in[41], and its four regimes of Stribeck friction curve are illustrated by reference [42] as in Fig. 1-22.

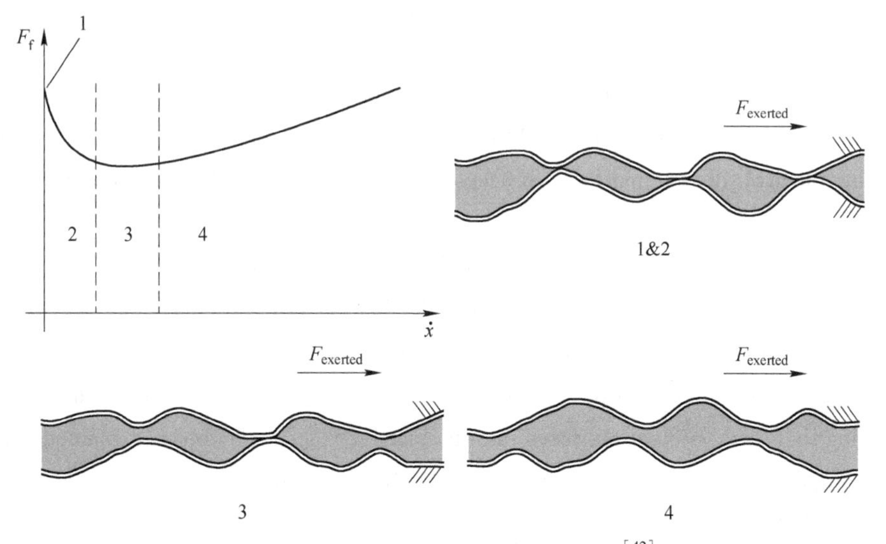

Fig. 1-22 Four regimes of Stribeck effect [42]

The Stribeck effect is related to the surface lubrication conditions, and the hydraulic fluid oil in the hydraulic system plays a role as lubrication. The four regimes are detailed interpreted in reference [43], they are static friction, boundary lubrication, partial fluid lubrication and full fluid lubrication. Friction in the stiction and boundary lubrication regimes are solid friction, and the velocity does not reach the level that the fluid film layer can be formed. Boundary lubrication can be regarded as a process of the shear in solids, and the solid shear strength is normally greater than fluids, thus the friction in the second regime is greater than that in the third regime [43].

The friction lag phenomenon is occurred around Stribeck effect as in Fig. 1-23, the friction force is larger when the velocity in acceleration state than that in deceleration state, a time lag of friction response is used to interpret it[44] and experimental test under lubricated condition is carried out to verify this phenomenon[45].

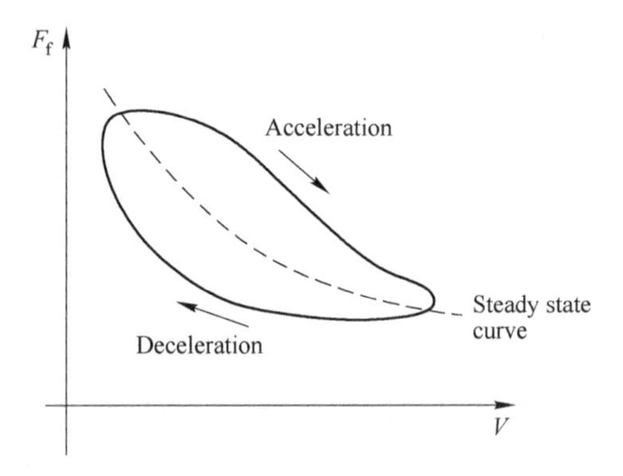

Fig. 1-23 Friction lag[46]

Such a phenomenon can be explained as a squeeze effect, in which time is required to form the new oil film thickness. With the lubrication conditions in Fig. 1-22, for a certain velocity, the friction force in acceleration state is larger than that in deceleration state.

Stick-slip motion is another nonlinearity in friction, it appears as oscillation during the load moves at a low velocity as in Fig. 1-24[47], and is not only affected by the condition of the contacting surfaces, but also the dynamics of the system.

The stick-slip motion can be regarded as the asperities springs repeat building up and breaking between contact surfaces as in Fig. 1-25[48], which is explained under macroscope scale and it is much more complicated under microscope level like in reference [49].

But such behaviour is mostly discussed under microscope level, which is not the

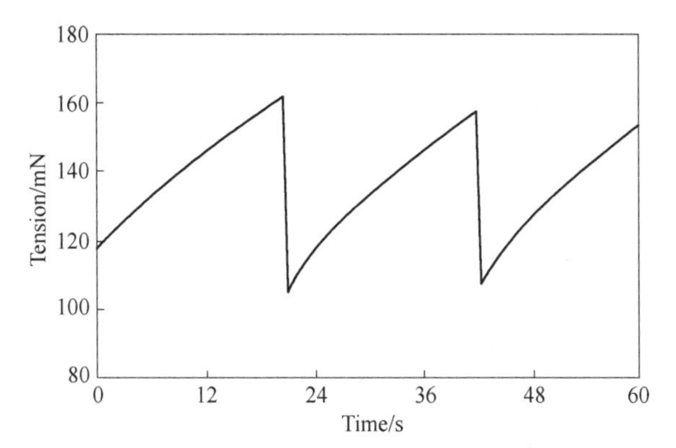

Fig. 1-24 Stick-slip motion of friction

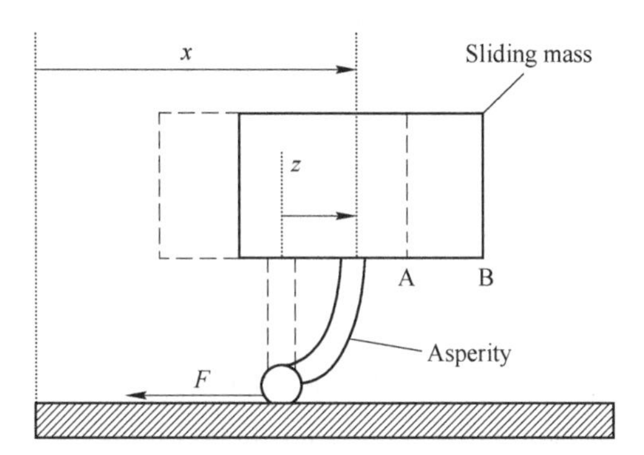

Fig. 1-25 Asperity between contact surfaces

concern of this research. Similarly, the pre-sliding phenomenon is also observed under microscope level and details can be found in reference [50], this phenomenon is also out of the research scope.

1.3.2 Friction model

Various friction models are proposed to simulate friction behaviours, they can be classified into two types, one is static models and the other is dynamic models. The major difference is that the static model does not include pre-sliding displacement which occurs at the contact interface. Picking a proper friction model for the corresponding design can help to improve simulation performance and system control. Hence, this section reviews several friction models and compares their differences for a better option for future research.

1.3.2.1 Static model

A typical static model is the classical friction model, which includes stiction friction, Coulomb friction and viscous friction. This model relates friction force with velocity and uses a discontinuity at zero velocity to explain the direction change of Coulomb friction. In this model, the friction can be any value between Coulomb friction limits at zero velocity, it is depicted in Fig. 1-26.

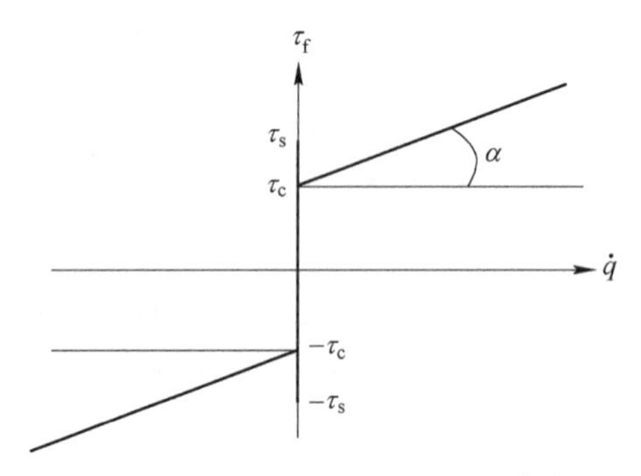

Fig. 1-26 Classical friction model[51]

This simple classical model has problems when the velocity crosses zero in simulation. Karnopp overcame this[52] by introducing a small velocity zone that the system stays as "stick" state, during which the stiction friction force exactly cancels the external driving force. After the driving force exceeds the "break free" force value, the system starts to accelerate. When the magnitude of velocity is outside the velocity zone, the system switches from "stick" to "slip". Though the model solved the zero-cross problem and improved computation efficiency, the complexity of the model increases when more masses are added to the system.

Tustin[53] proposed a model that predicts friction forces when the velocity is closed to zero, named as Tustin model. The exponential model is implemented in a general form reference [54].

These friction models are typical static friction models, and they do not include pre-sliding displacement compared to a dynamic model, such a phenomenon is reviewed in the next section.

1.3.2.2 Dynamic model

One of the first dynamic models is created by references [55]-[56], the Dahl model is based on experimental work of ball bearings. The pre-sliding displacement is interpreted as the elastic deformation of surface asperities as in Fig. 1-27, which named as Dahl effect. The friction force during the pre-sliding can be depicted as a function of displacement that in the range of 1-50 μm. Though the Dahl model is simple to use, it does not include stiction or Stribeck effect.

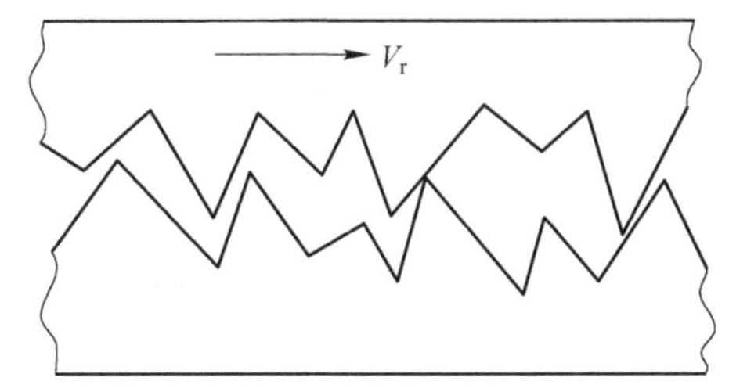

Fig. 1-27 Surface asperities deformation[57]

The bristle model is constructed[57], which modeled the contacting asperities on the surface as rigid bristles on one surface and elastic ones on the other as in Fig. 1-28. The sum of the restoring forces acting on the elastic bristles is regarded as friction force. z stands for the average deflection of the bristles.

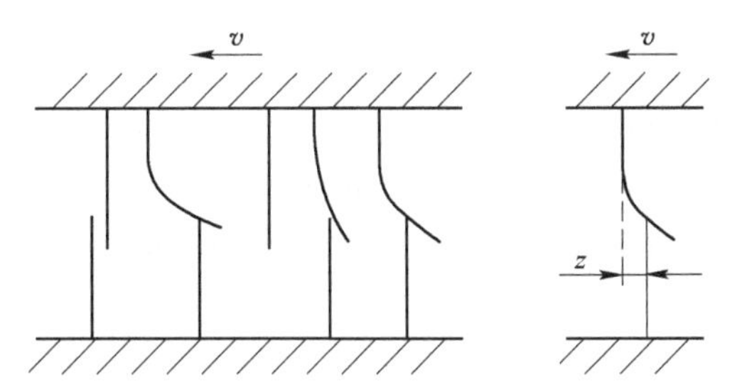

Fig. 1-28 Bristle model[57]

As the bristle model needs to calculate numbers of bristles' behaviour, it is computationally inefficient. The reset integrator model[57] is built by Haessig and Friedland to reduce the computer time required by the Bristle model, meanwhile, retaining its accuracy to describe the stick-slip phenomenon. This model utilised the average deflection z to state the system in stick or slip phase. Both models predict the stick-slip friction, but Stribeck effect and friction lag phenomenon are not included.

Based on the bristle model, the LuGre model is put forward[58], named after the two universities, Lund and Grenoble. This model is simple and includes most of the friction phenomenon, especially when the system in the low-velocity state. It captures hysteresis-like behaviours containing friction lag, spring-like characteristics in stiction and stick-slip behaviour during velocity reversals[59].

There exist some other friction models that improved the LuGre model, like the Generalized Maxwell-slip model[60], Leuven integrated friction model[61-62], etc. Most of them focus on the pre-sliding and hysteretic phenomenon, which occurs in a very small displacement range whose unit is μm. However, for most hydraulic applications, they are not able to operate in such a small displacement range. So, the LuGre friction model is reviewed in detail in this section.

The asperities junctions that used in the LuGre model are considered as bristles that illustrated as in Fig. 1-28, the average deflection of bristles is developed with the below equation:

$$\frac{\mathrm{d}z}{\mathrm{d}t} = v - \sigma_0 \frac{z}{g(v)} |v| \tag{1-7}$$

where v is the relative velocity, m/s; $g(v)$ is used to describe the Stribeck effect and the overall friction force F_f is given as:

$$F_f = \sigma_0 z + \sigma_1 \frac{\mathrm{d}z}{\mathrm{d}t} + \sigma_2 v \tag{1-8}$$

where σ_0 is the average stiffness of the bristles; σ_1 is a micro-viscous coefficient which is equivalent to a damping coefficient; $\sigma_2 v$ is the viscous friction.

A parameterisation expression for $g(v)$ can be written as below to describe the Stribeck effect:

$$g(v) = F_c + (F_s - F_c)\mathrm{e}^{-(v/v_s)^n} \tag{1-9}$$

where F_c is the Coulomb friction; F_s is the stiction friction and v_s is the Stribeck velocity; n is an arbitrary constant that is related to geometry and usually set to unity.

So that this model is characterized by these six parameters, some values can be identified when the velocity is under a steady state.

So that when the velocity is under a steady state, the average bristle deflection z becomes:

$$z_{ss} = \frac{v}{|v|} g(v) = g(v)\,\mathrm{sgn}(v) \tag{1-10}$$

where

$$\mathrm{sgn}(v) = \begin{cases} 1 & \text{if } v > 0 \\ 0 & \text{if } v = 0 \\ -1 & \text{if } v < 0 \end{cases} \tag{1-11}$$

Furthermore, the steady states friction force should be:

$$F_{rss} = g(v) + \sigma_2 v \tag{1-12}$$

After rearranged the equations, the steady state friction equation is:

$$F_{rss} = F_c + (F_s - F_c)e^{-(v/v_s)^n} + \sigma_2 v \tag{1-13}$$

The steady state friction force F_{rss} and the velocity v can be measured, the stiction friction force F_s and Coulomb friction force F_c can be observed and measured when the system switches from stationary to a constant velocity, n is an appropriate exponent. So that the viscous coefficient σ_2 can be identified. As the bristle average deflection is an imaginary parameter, the parameter σ_0 and σ_1 is not physically measurable.

1.4 Concluding remarks

In this chapter, numbers of energy efficiency solutions for the hydraulic system are reviewed in the first section, most of these solutions can be classified into two major types, the energy-efficient valve-controlled system with fixed supply pressure and the pump directly controlled system. The pump directly controlled system has more energy efficiency potential with the following advantages:

(1) Energy efficient: the throttle losses in the control valve are minimised.

(2) Cost-friendly: a servo motor and a fixed displacement pump are usually cheaper, and they are the power source with only two components.

(3) Space saver: The design is compact as there is no need for pressure compensators.

The pump-controlled system has its disadvantage, which is mainly about the pressure stability performance under some certain conditions, which makes such design more suitable for less dynamic performance hunger devices.

A component linking method is reviewed for simulation methodology. It is able to observe almost all the performance of each component during system operation simulation, which is helpful for further research on the system. Another simplified method is the pump-controlled system modeled by a state space, which is suitable for controller design purposes.

Various friction models are reviewed. Therefore, the LuGre model is the most suitable one for hydraulics, and it captures almost all the overall friction phenomenon without being overly complicated. There exist more detailed friction models, but they are more likely to focus on the pre-sliding region, which is out of this research scope. However, the LuGre friction model only relates the friction force to velocity and displacement, but the friction in hydraulic is affected by various conditions, so some improvement can be achieved.

So that the concept of the pump-controlled asymmetric cylinder is chosen for energy-efficient purpose, the LuGre friction model will be added to the future simulation for more nonlinear behaviours observations.

Chapter 2 Component Linking Modeling Method

The original component linking method modeling is modeled in Fortran by Leaney in 1986, to achieve a further nonlinearities analysis in Simulink, the component linking modeling method needs to be validated in Simulink in the first place. After the method is validated, further nonlinearities research when the valve in the underlap region is carried out in Simulink in an open loop test, an analytical solution when the control valve in underlap region is implemented and the computing time efficiencies are compared.

2.1 Moog valve controlled asymmetric cylinder system

A Moog four-way valve-controlled asymmetric cylinder system is constructed and tested by Leaney[63]. This test rig is presented as a position-controlled asymmetric drive in conjunction with a plain slideway load. The plain slideway contains a slideway table that is used on a Moog Hydrapoint milling machine. Pressure transducers are integrated to measure its service line pressures, potentiometer and a velocity transducer is applied to measure the position and velocity of cylinder respectively. Its system scheme and pictures are displayed in Fig. 2-1.

Leaney[63] in his work modeled this system in Fortran, the equations he used are reviewed in this chapter and they will be utilised to reproduce the model in Simulink.

2.1.1 System components

This section presents the parameters of the major components in the Moog valve controlled asymmetric cylinder drive system, including the Moog four-way valve, asymmetric cylinder, fluids and power source.

The four-way symmetric valve Moog series 76 model 102[64] in Fig. 2-2 was selected

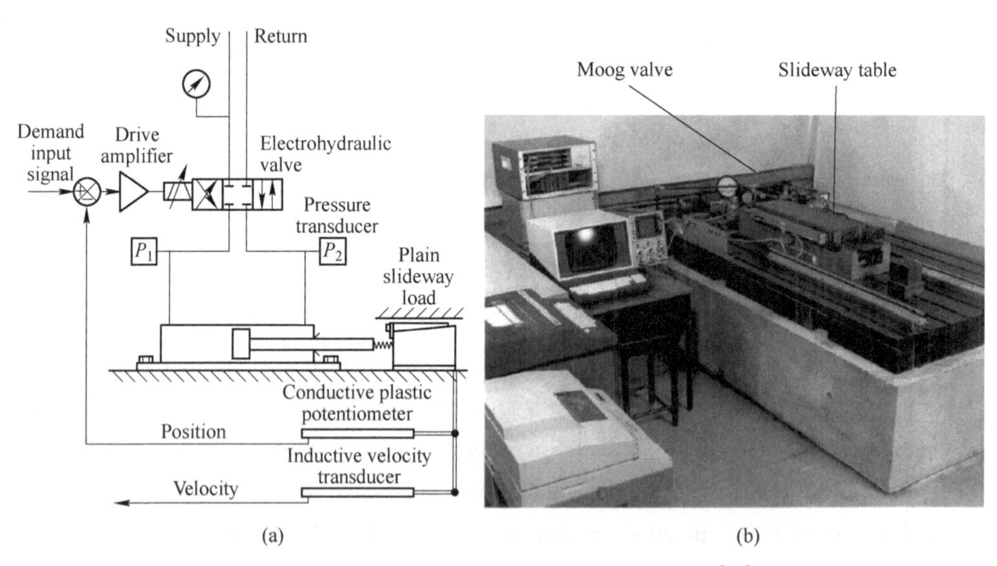

(a) (b)

Fig. 2-1 Moog valve controlled asymmetric system[63]

(a) Schematic diagram; (b) Test device overview

for this test rig, though the valve is claimed to be zerolapped valve, the results from blocked port tests for pressure gain and null leakage flow[63] indicates that there is an equivalent underlap region in this valve.

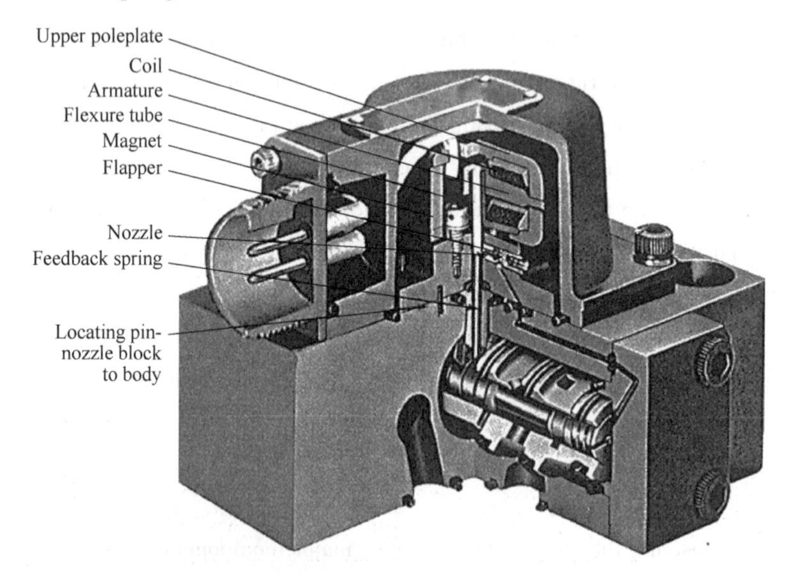

Fig. 2-2 Moog series 76 model 102 valve[64]

The valve block port test indicates an equivalent underlap region of ±0.0045 mm which is ±1.2% of the maximum spool stroke[63]. As the valve information from its data sheet[64], the valve parameters can be concluded in Table 2-1.

Table 2-1 Moog four-way valve parameters

Valve stroke/mm	±0.375
Underlap region/mm	±0.0045
Spool diameter/mm	7.938
Fraction of fully annular ports	1
Time constant/s	0.004

The asymmetric cylinder with port number 88533-002 is used as the only actuator of the test device, and its parameters and hydraulic oil fluid properties are listed in Table 2-2.

Table 2-2 Asymmetric cylinder geometry & fluid properties

Cylinder stroke/m	0.52
Piston area/cm^2	11.43
Rod area/cm^2	5.8
Oil density/$kg \cdot m^{-3}$	870
Effective oil modulus/MPa	700
Supply pressure/MPa	7

Normally, the oil modulus should be much higher than 700 MPa, the lower value may be caused by air trapped in the oil, which will lead to a decrease in oil bulking modulus. This section collects the parameters that will be utilised in the component linking method simulation in Simulink, which will be carried out in the next section.

2.1.2 Component linking model in Simulink

The model is composed of three parts valve model, actuator model and load model. In the valve model, the valve has three states as in Fig. 2-3, the judgment of the state is required during operation.

The valve receives signal current and its spool moves to the target position, due to the valve dynamic equation in section 1.2.1.1 and its parameters, the valve dynamic can be indicated as:

$$\frac{X_s(s)}{I(s)} = \frac{K_s}{\tau_s s + 1} = \frac{0.03}{0.004s + 1} \tag{2-1}$$

The saturation block in Fig. 2-4 is to limit the spool displacement to its physical maximum value.

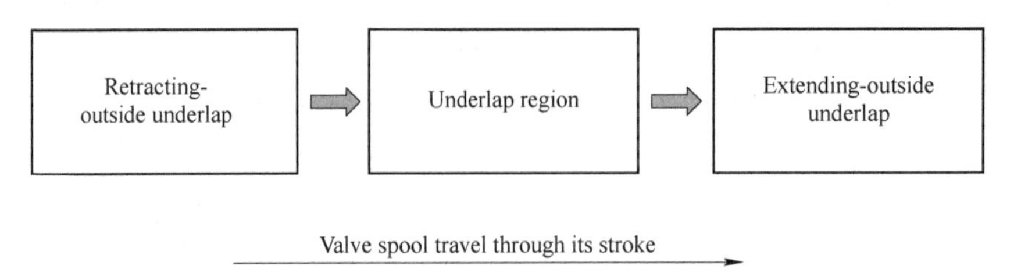

Fig. 2-3　Three valve states

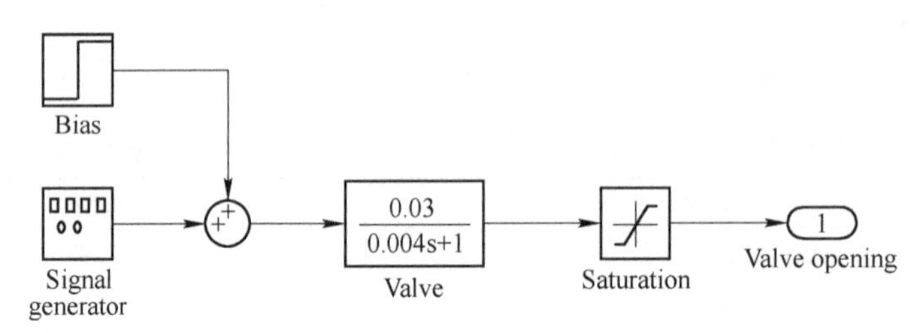

Fig. 2-4　Moog four-way valve dynamics, valve model in Simulink

With the calculated spool displacement x_s, the valve opening area A_{v1} and A_{v2} can be revealed with following equations:

$$\begin{cases} A_{v1} = \begin{cases} D\pi(0.0000045 + x_s), & (0.0000045 + x_s) > 0 \\ 0, & (0.0000045 + x_s) \leqslant 0 \end{cases} \\ A_{v2} = \begin{cases} D\pi(0.0000045 - x_s), & (0.0000045 - x_s) > 0 \\ 0, & (0.0000045 - x_s) \leqslant 0 \end{cases} \end{cases} \tag{2-2}$$

The valve opening area is described in the physical layout in Fig. 2-5, equivalent underlap is depicted.

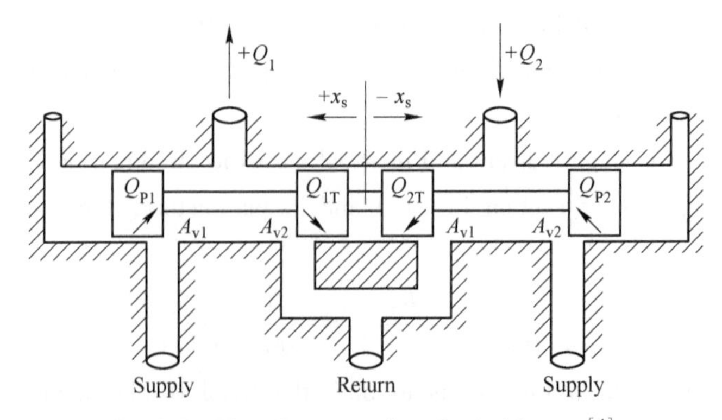

Fig. 2-5　Moog four-way valve physical layout [4]

Eq. (2-2) above point out how the valve opening area changes when the valve in different cylinder states, when both areas A_{v1} and A_{v2} are none zero, the valve is inside the underlapped region. When the valve is operating outside the underlap, $A_{v1} = 0$ & $A_{v2} > 0$ or $A_{v1} > 0$ & $A_{v2} = 0$.

Compressibility flow Q_{CE} can be obtained by equation below, which is used to calculate the effective load pressure P_{LE} by Eq. (2-3).

$$Q_{CE} = Q_{LE} - Q_{DE} - Q_{KE} \tag{2-3}$$

Substitute P_{LE} and X factor into equations and to obtain the chamber pressures p_1 and p_2 [32].

These processes in Simulink are presented in Fig. 2-6.

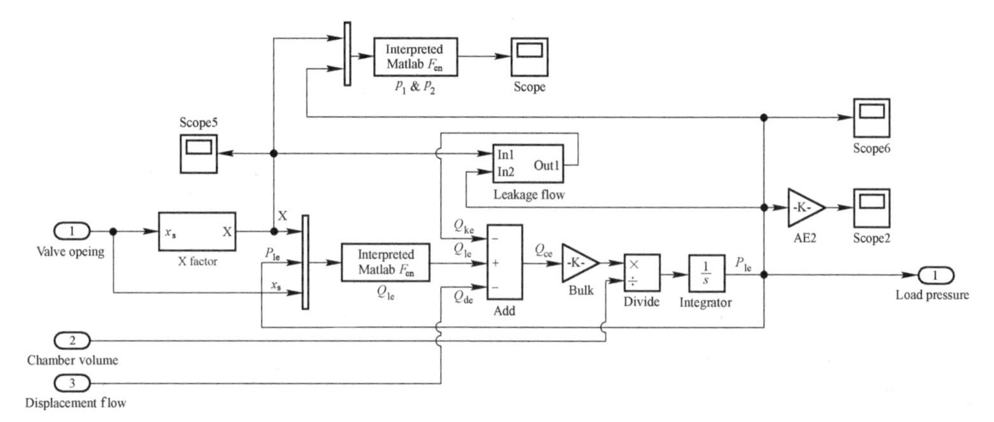

Fig. 2-6 Actuator model in Simulink

The load model structure is simple. It accepts the force generated by cylinder and considers the influence of friction, then outputs the net force to the mass. At last, use integrator to obtain velocity and displacement from acceleration, the Dormand-Prince method[65] is utilised for integration operation in Simulink. These processes are achieved as in Fig. 2-7.

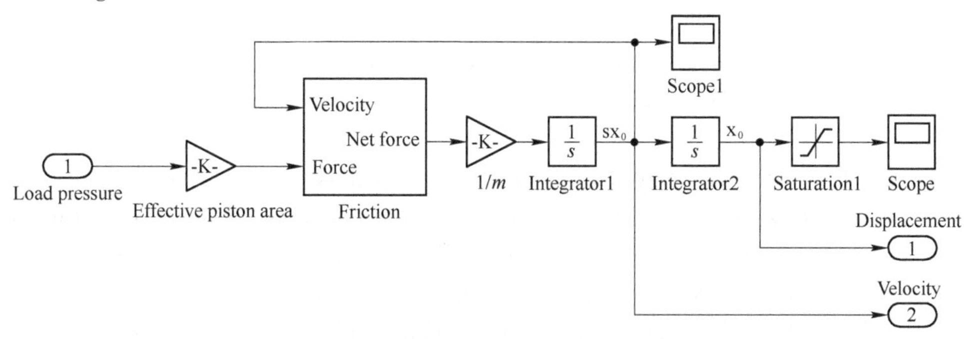

Fig. 2-7 The load model in Simulink

Next section combines all three models to create an overall model, and its simulation results are compared with published results from reference [63].

2.1.3 System model in Simulink

This section collects all the simulation results from the whole model in Fig. 2-8, to compare with the published test results[32] to validate the component linking method in Simulink.

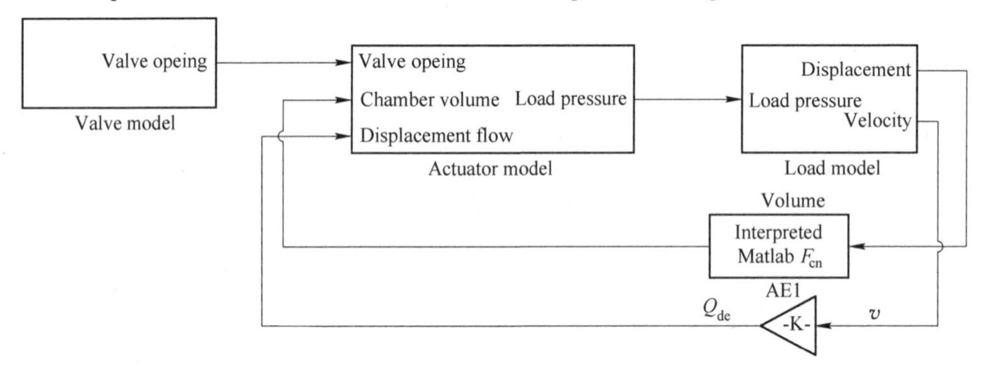

Fig. 2-8 Component linking method in Simulink

The overall model block diagram is depicted in Fig. 2-9.

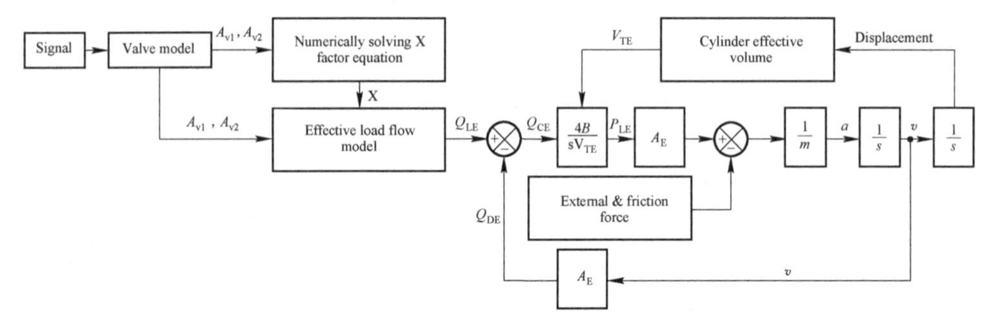

Fig. 2-9 Block diagram of the numerical solving method for valve controlled asymmetric cylinder drive system

This method is used to model an open loop test of a four-way valve controlled an asymmetric cylinder drive system, square wave input signal controls the valve opening area and the cylinder will follow to change its state between the extending and retracting. However, as the symmetric ported valve controls the asymmetric cylinder, if the input square wave is symmetric, the asymmetric cylinder will gradually retract to its end due to it retracts faster than extending. So that, a bias is added to the signal to create a balanced cylinder movement as in Fig. 2-4. The Simulink modeling results are listed in Fig. 2-10 and Fig. 2-11, and then compared with published test results[63].

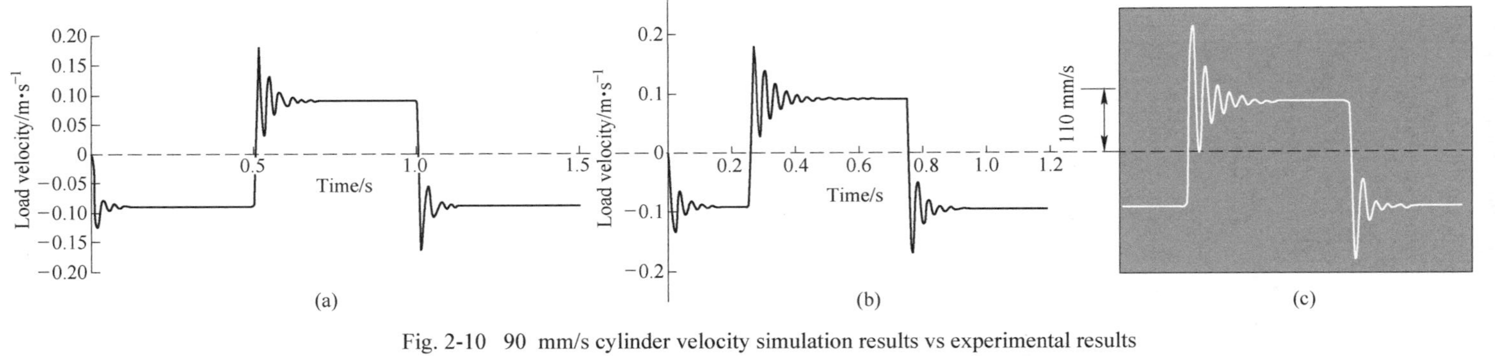

Fig. 2-10 90 mm/s cylinder velocity simulation results vs experimental results

(a)Simulink results; (b)FORTRAN simulation results; (c)Experimental results

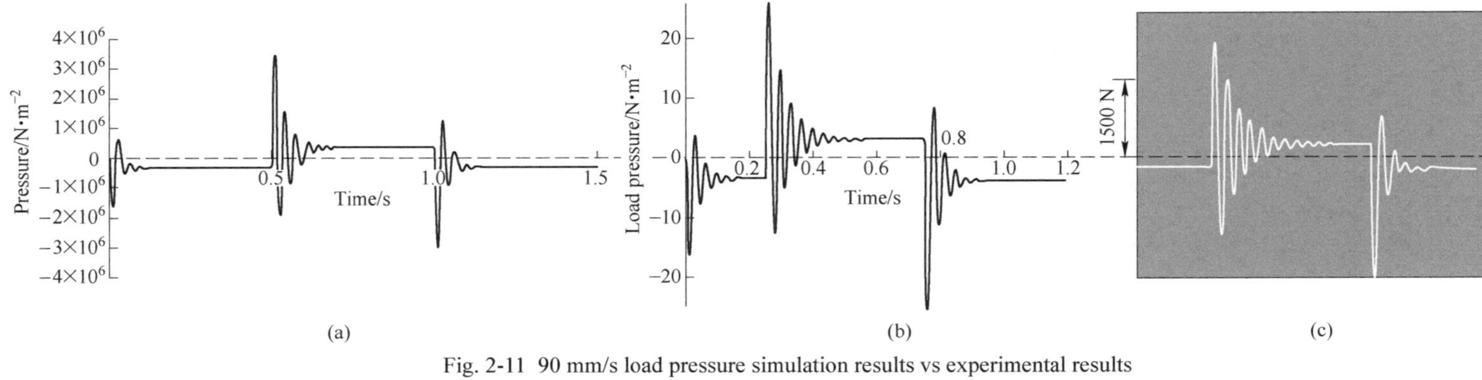

Fig. 2-11 90 mm/s load pressure simulation results vs experimental results

(a)Simulink results; (b)FORTRAN simulation results; (c)Experimental results

The modeling results from Simulink shows consistency with the modeling results from Fortran and experimental results[63]. It indicates that the component linking method is validated in Simulink and can be utilised for further nonlinearities research of the asymmetric cylinder drive system.

2.2 Concluding remarks

The component linking method in Chapter 1 is validated in Matlab Simulink, and its simulation results show consistency with the published results.

This chapter validated and improved the component linking method, which will be utilised in a different asymmetric cylinder drive system in Chapter 3.

Chapter 3 Hybrid Pump Controlled Asymmetric Cylinder System

As reviewed in Chapter 1, the valve-controlled asymmetric cylinder drive system has good controllability, but its energy efficiency is an inevitable issue due to its throttle losses and the requirement of constant supply pressure. The pump-controlled asymmetric cylinder drive system has much better performance in energy conservation aspect, but its stability during operation is a concern under some certain conditions. To combine the advantages of both types of asymmetric cylinder drive system, an open loop hybrid pump-controlled asymmetric cylinder system is put forward and implemented as in Fig. 3-1.

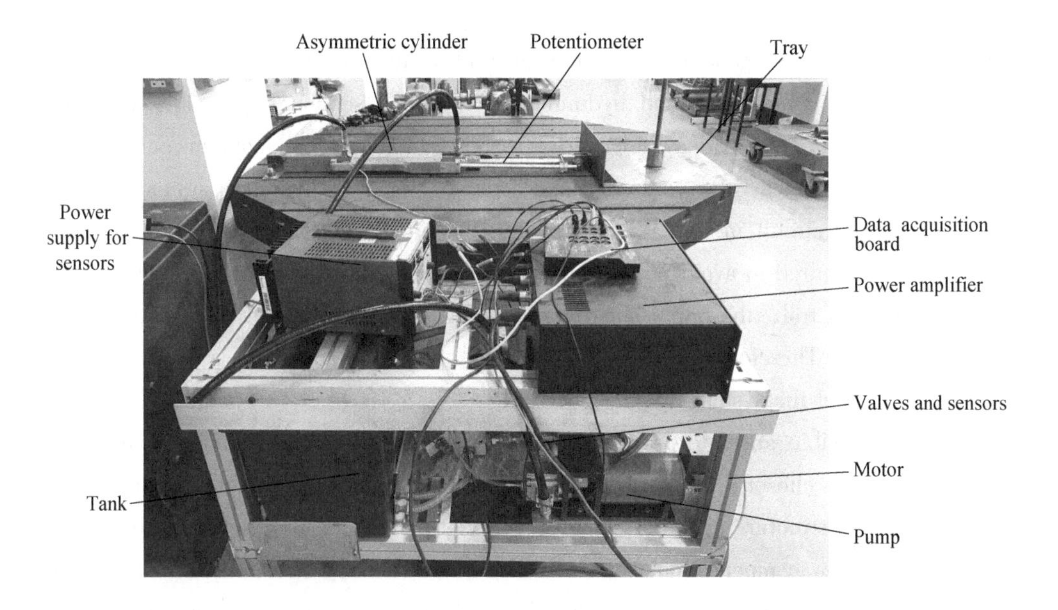

Fig. 3-1 Hybrid pump controlled asymmetric cylinder drive system

3.1 Design blueprints

The pump-controlled asymmetric cylinder system consumes only 11% of the energy required by a valve-controlled system to perform the same task[66]. Load-sensing circuit improves the energy efficiency of a hydraulic excavating machine, but the throttling losses are still occupying at least 35% of the power input[67]. Similar situations can be found in energy-efficient solutions for the valve-controlled hydraulic system.

A pump-controlled hydraulic cylinder drive system can significantly reduce the power consumption compared to a valve-controlled hydraulic system. For a symmetric cylinder system, the pump-controlled system shows acceptable controllability due to the balanced flow delivered into cylinder chambers[19]. But when the actuator is an asymmetric cylinder, velocity oscillations are observed under some certain condition [25]. However, a purely pump-controlled hydraulic system has a low stiffness when holding the load. When the cylinder is commanded to maintain a position and the pump is stationary, loop stiffness is required when the load changes in the commanded position[68].

In at least 80% of the electro-hydraulic applications, the asymmetric cylinders are used. Without loss of generalisation and to find a middle place between a valve-controlled and a pump-controlled hydraulic system, this section implements a hybrid pump-controlled asymmetric cylinder drive system circuit, which contains advantages and disadvantages of a valve-controlled and a pump-controlled hydraulic system.

The proposed circuit in Fig. 3-2 is based on the pump-controlled concept, but flow regulation is required to avoid possible instabilities during operation. As the requirement of the system control, the pump must be a servo-type pump (the only controllable unit in the system). Therefore, a bidirectional gear pump coupling with a servomotor is utilised to control the system flow rate.

To avoid possible unstable problems, an open circuit design including a directional four-way valve is chosen, and a needle valve for flow regulation is required in the return line. Due to the structure of the asymmetric cylinder drive, the flow will be unbalanced if there is no flow compensation mechanism involved. Therefore, two check valves that are connected to the oil tank are introduced into the system to avoid cavitation.

To minimise the throttle losses, the four-way valve should be oversized. Due to the system requirements of low cost and simplicity, a pilot-shifted four-way valve is chosen

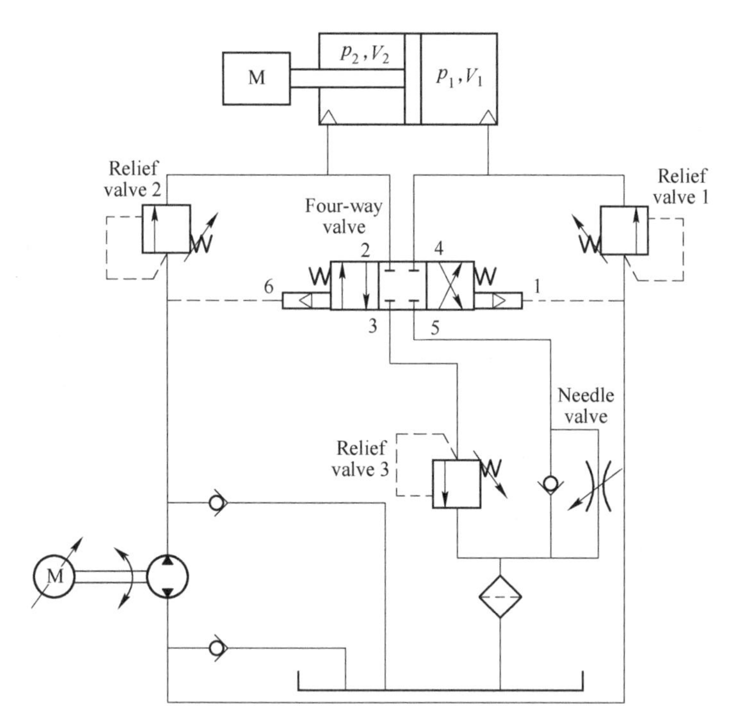

Fig. 3-2 Hybrid pump controlled asymmetric cylinder drive

1, 6—The two ends of the valve core of a three-way four-way valve;

2-5—The valve hole of a three-way four-way valve

to control the cylinder direction of motion. Relief valve 1 and 2 are placed in the service line to provide the required pilot pressure to drive the four-way valve, their pressure settings are set to the minimum values required to shift the four-way valve. The needle valve is manually adjusted to regulate the flow. The relief valve 3 is used as a safety valve with 20 MPa pressure setting. The filter is to remove the particles in the oil on a continuous basis.

When the pump runs clockwise, the system drives the asymmetric cylinder extending as in Fig. 3-3 (a), runs counter clockwise to drive the asymmetric cylinder retracting as in Fig. 3-3 (b).

The dotted arrow in Fig. 3-3 indicates the fluid flow direction. This design combines the valve-controlled and pump-controlled asymmetric cylinder drive system, the pump first charges the pipeline to pilot shift the four-way valve to create a pathway to the tank. The four-way valve is over-sizing for the system, which is to reduce the throttle losses as much as possible, the pre-set relief valves next to it is set to a value that is able to fully shift the four-way valve. The needle valve in the return line is to add

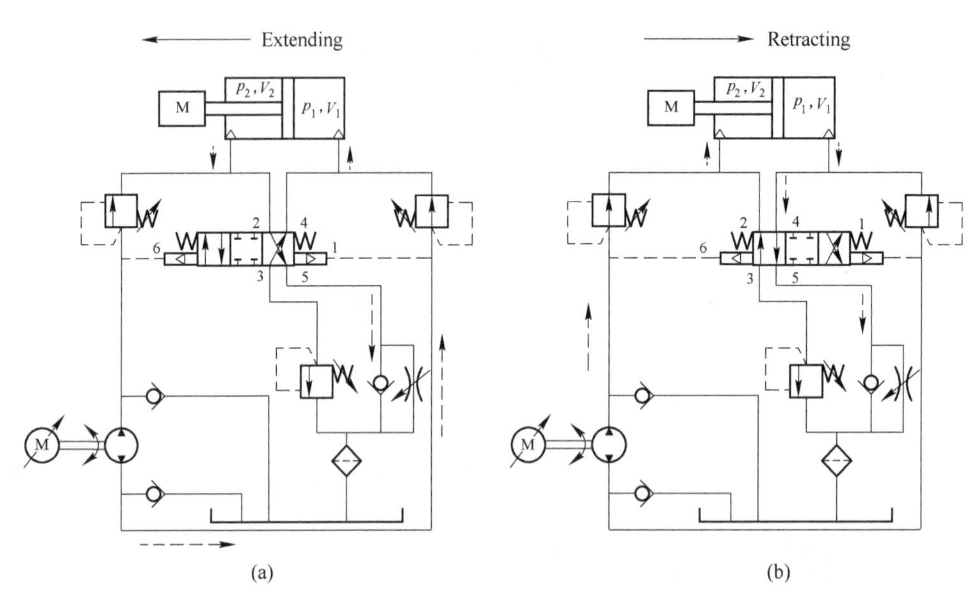

Fig. 3-3 Proposed hybrid pump controlled asymmetric cylinder drive system in extending (a) and retracting (b) state

stiffness to the system by regulating the flow rate. The valve opening of the needle valve is set to a proper value to achieve a balance between throttle losses and stability. Its operating processes are illustrated in Fig. 3-4.

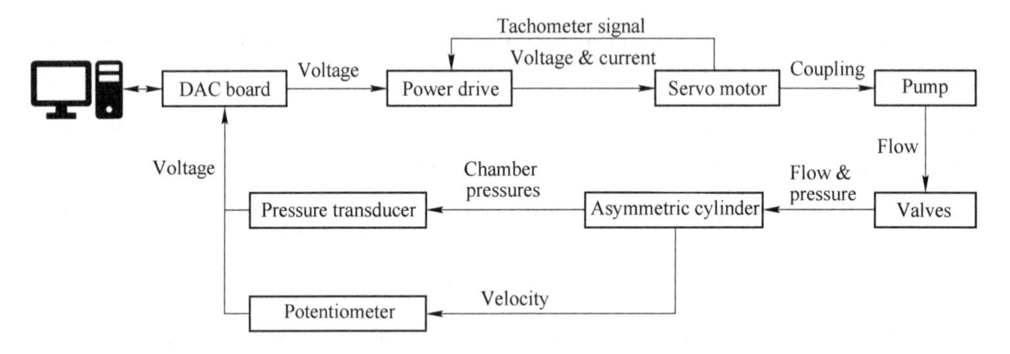

Fig. 3-4 Operation flow chart for pump controlled asymmetric cylinder drive system

The choice of a pilot-shifted four-way valve is based on the target of this research. A conventional four-way valve in hydraulic applications plays both roles of directional control and flow control, but the pilot-shifted four-way valve in Fig. 3-2 is oversizing and only used for directional control. As the pump must charge the pilot pressure, the pilot shifted valve does not fast response as a servo valve. In theory, this design circuit is suitable for less dynamically demanding hydraulic applications. How this system performs will be modelled and tested in later chapters. Its components and parameters are investigated in the next section.

3.2 System components and parameters

The hybrid pump-controlled asymmetric cylinder drive system circuit has its potential advantages and disadvantages. To simulate and analyse this device design, its components must be modelled and their parameters should be obtained in the first place. To ensure the component linking modeling method can be utilised for simulation, system components' parameters are collected and measured in this section.

3.2.1 Servomotor

A DC servomotor is the power source of the system, the model number of this SEM motor is MT30R4-25. The motor accepts voltage from the power amplifier to run at a certain speed, the internal tachometer sends the motor speed information back as the form of voltage. Its major parameters are listed in Table 3-1.

Table 3-1 The specifications of MT30R4-25 servomotor[69]

Parameter	Value
Voltage gradient (at 1000 r/min)/V	25
Max terminal Voltage/V	100
Max speed/r · min^{-1}	4000
Cont. stall torque/N · m	3.2
Cont. stall current/A	13.3
Peak torque/N · m	18
Peak current/A	86
Torque constant/N · m · A^{-1}	0.24
Voltage constant/V · s · rad^{-1}	0.24
Armature resistance/Ω	0.4
Armature inductance/mH	2.2
Mechanical time constant/ms	14
Motor weight/kg	8.3

This servo motor is coupled with a bidirectional gear pump to deliver fluid into the system.

3.2.2 Pump

The pump applied in the system is a bi-directional gear pump from HPI, which is coupled with the servo motor and one extra port is connected to the oil tank. The pump does not only deliver fluid to the system, but is also able to compensate the unbalanced flow rate and prevent cavitation. Its specifications are listed in Table 3-2.

Table 3-2 The specifications of the HPI pump[70]

Parameter	Value
Capacity/mL · r^{-1}	2.05
Peak pressure/MPa	30
Max working pressure/MPa	25.5
Max speed/r · min^{-1}	8000
Nominal flow (at 1500 r/min)/L · min^{-1}	3.07
Nominal flow at max speed/L · min^{-1}	16.4
Rotation direction	Bi-direction

The key parameter of the pump is the flow rate, but due to the circuit design and manufacture tolerance, the actual nominal flow rate may have some bias compared to its datasheet.

3.2.3 Power supply and motor drive

The matched motor drive is Microspeed 110 series from AXOR, which is the controller of the motor. The power supply accepts 240 VAC from the power socket and transfers to 100 VAC to power the Microspeed controller, then the motor drive outputs demanding VDC to the motor to control its speed[71]. Feedback control is utilised in this drive, after a gain is set (based on the tachometer constant of the servo motor), the drive ensures the motor speed is consistency with the reference signal.

3.2.4 Valves

There are three types of valves in this hybrid pump-controlled asymmetric cylinder drive system, including a model DDDC four-way pilot shifted valve, a model RBAE pressure relief valve and a model NCCC needle valve. These valves are all from SUN Hydraulics.

The model DDDC four-way valve is a spring-centred oversizing valve, its circuit view is described in Fig. 3-5.

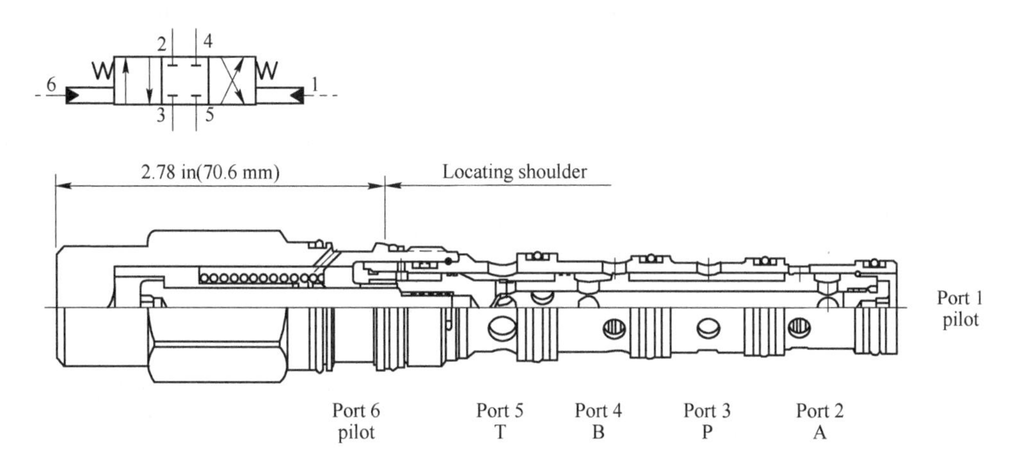

Fig. 3-5 Circuit and section view of DDDC four-way valve[72]

It is a three-position, four-way directional cartridge with six ports, all ports can accept 35 MPa pressure, its technique data is listed in Table 3-3.

Table 3-3 The specifications of the DDDC four-way directional valve[72]

Parameter	Value
Capacity/L · min^{-1}	75.71
Minimum pilot shift pressure/MPa	1
Max working pressure/MPa	35
Weight/kg	0.5

Its factory performance curves are listed in Fig. 3-6.

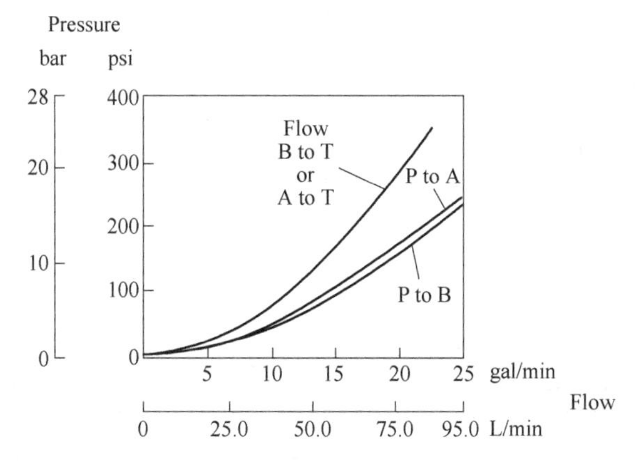

Fig. 3-6 Performance curves of the DDDC four-way valve [72]

(1 bar = 14.5 psi = 0.1 MPa)

The capacity of this valve is much larger than the other components capacity, this choice is to reduce the throttle losses to the minimum. The performance curve in Fig. 3-6 indicates this DDDC valve is not a symmetric ported valve, the port 3 is larger than port 5 but port 2 and port 4 are equal. This design will affect the nonlinear behaviours of the hybrid pump-controlled system, and the details will be discussed in Chapter 5.

Three RBAE pressure relief valves are utilised in this system, two of them are placed next to the DDDC valve, their inlet ports are connected to the four-way valve pilot ports as in Fig. 3-2. This arrangement is to ensure the four-way valve is shifted before the hydraulic fluid flows into cylinder chambers. The pressure setting of the two RBAE valves is adjusted to the minimum shift pilot pressure of the DDDC valve, the third one is used as a safety valve. Its circuit and section view are listed in Fig. 3-7.

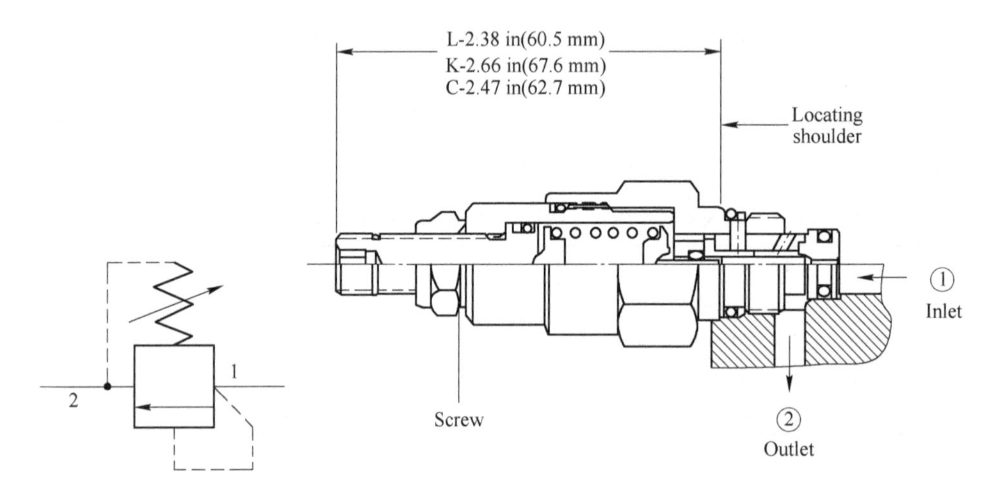

Fig. 3-7 RBAE pressure relief valve[73]

Its technique data is listed in Table 3-4.

Table 3-4 The specifications of RBAE relief valve[73]

Parameter	Value
Capacity/L · min^{-1}	9. 5
Response time/ms	2
Max working pressure/MPa	35
Adjust range/MPa	0. 17 ~ 21
Weight/kg	0. 13

The pressure setting is adjusted by turning the screw that locates on the top of the valve. Due to the design, the backpressure at the outlet is directly additive to the pressure at the inlet. Its performance curves are depicted in Fig. 3-8.

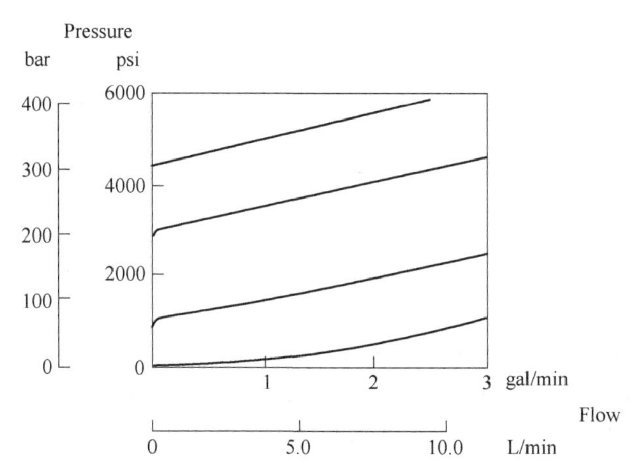

Fig. 3-8 RBAE pressure relief valve performance curves [73]
(1 bar = 14.5 psi = 0.1 MPa)

A needle valve is placed at the return line of the system, which is used to regulate the flow back to the tank. This arrangement is to increase the stiffness of the system, which increases the stability during operation. Its circuit and section view are listed in Fig. 3-9.

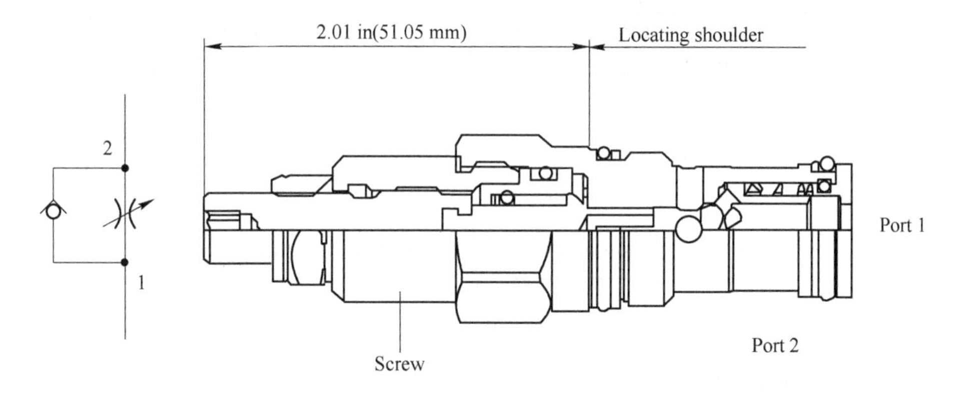

Fig. 3-9 NCCC needle valve [74]

This needle valve is equipped with a reverse flow check component. The needle valve is adjustable from fully closed to maximum orifice diameter by turning the screw on top. Its technique data is listed in Table 3-5.

Table 3-5 NCCC needle valve specification[74]

Parameter	Value
Capacity/L · min^{-1}	9.5
Adjustment/turn	5
Max working pressure/MPa	35
Weight/kg	0.15

3.2.5 Actuator

The hydraulic actuator in this hybrid pump-controlled system is an asymmetric cylinder drive from AIR POWER & HYDRAULICS. It is C10 cylinder 18 series with a stroke of 300 mm, 18 mm rod diameter and 32 mm piston diameter. Its section view is depicted in Fig. 3-10, and a H shape seal is applied in this cylinder.

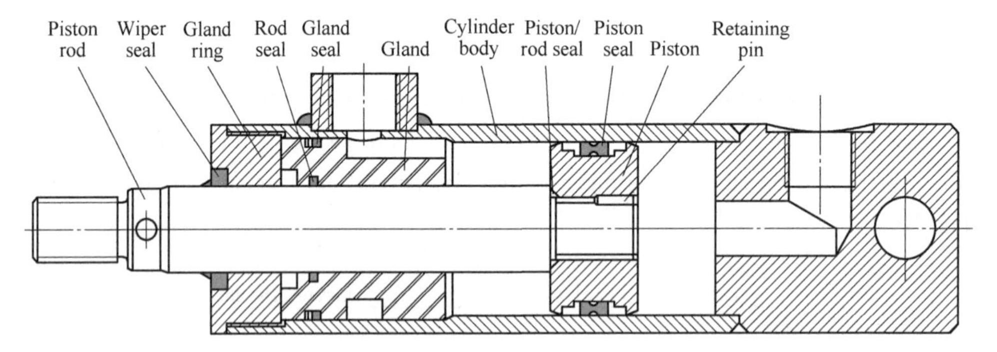

Fig. 3-10 Section view of the C10 series 18 model asymmetric cylinder[75]

Its technique data is listed in Table 3-6.

Table 3-6 C10 series 18 asymmetric cylinder specifications[75]

Parameter	Value
Piston diameter/mm	31.75
Rod diameter/mm	18
Max working pressure/MPa	35
Stroke/mm	300
Piston area/cm^2	8.042
Annular area/cm^2	5.50
Maximum speed/mm · s^{-1}	500

3.2.6 Sensors

The sensors in this hybrid pump-controlled asymmetric cylinder drive system include a potentiometer and four pressure transducers. The potentiometer is attached to the cylinder to record the cylinder displacement. Two pressure transducers are connected to the cylinder chambers to measure the cylinder chamber pressures.

The potentiometer used is model CFL 300 from Sakae as depicted in Fig. 3-11 with a stroke of 300 mm. It is a basic potentiometer. 10 volts power supply is applied on terminal 1 and 3, and slider 2 provides voltage change which indicates the motion of the cylinder.

Fig. 3-11 Circuit of CFL 300 potentiometer from Sakae[76]

The four pressure transducers are the same, they are PXM309-350G pressure transducers from Omega. This PXM309-350G model is able to measure up to 35 MPa gauge pressure. The red and black cables are connected to the power supply, and the white is output the measured signal. The technique data of the transducer is listed in Table 3-7.

Table 3-7 Omega PXM309-350G pressure transducer[77]

Parameter	Value
Power supply/V	15 ~ 30
Output/V	0~10
Measurable pressure/MPa	0 ~ 35
Response time/ms	1
Bandwidth/Hz	1000
Accuracy/%	±0. 25

3.2.7 Signal processing platform

The hybrid pump-controlled asymmetric cylinder drive system utilises a Quanser USB Q8 data acquisition board to process the signal in real time. The board can process digital and analogue signal. As for the requirements of this hybrid pump-controlled system, only the analogue part is utilised.

This DAC board collects all the information from sensors and transfers them to the host PC. After the PC processes these signals, the board accepts the command from PC host and output analogue signal to target components. A summary of signal arrangement is listed in Table 3-8.

Table 3-8 Signal ports arrangement on the DAC board[78]

Signal	Component	Form	Pin type
Motor speed command	Motor drive	Voltage	Analog output
Motor speed feedback	Motor tachometer	Voltage	Analog input
Power supply	Potentiometer	Voltage	Analog output
Cylinder displacement	Potentiometer	Voltage	Analog input
Chamber pressures	Transducers	Voltage	Analog input

3.3 Concluding remarks

This chapter puts forward an open loop design circuit of a hybrid pump-controlled asymmetric cylinder drive system, targeting at combining the advantages of a valve-controlled and a pump-controlled asymmetric cylinder system, meanwhile, achieving the energy saving purpose.

Due to the pilot-shifted four-way valve being shifted by the pressure in the pipeline and there being a charging process when pump changes its direction, this hybrid design is not able to perform as fast response as a servo valve controlled system. As this research is not focused on dynamically demanding applications like in aerospace and military, the choice of this type of four-way valve should be acceptable. This oversizing four-way valve is applied to reduce the throttle losses to the minimum. But the needle valve in the return line is added to increase the stiffness of the system during operation to reduce the velocity oscillations possibility in the hybrid pump-controlled asymmetric cylinder system.

The system components are reviewed, and their parameters are collected for further simulation in Chapter 5. The design purpose will be tested and validated in Chapter 6.

Chapter 4 A New Friction Model Based on LuGre Model

The LuGre model is a friction model that simulates friction behaviours based on velocity only[43]. However, for the hydraulic applications, the major part of the overall system friction force is from cylinder seal and this part is not only affected by the velocity [8]. This chapter investigates other factors that will affect the friction other than the velocity, and implements a new friction model on current LuGre model to better capture the friction behaviour in hydraulics, including pressure term, acceleration term and velocity term.

4.1 Implement of the new friction model

As reviewed in Chapter 1, the LuGre model captures almost all the friction phenomenon without a complex model structure. The LuGre model composes three equations as follows, where v is the relative velocity of two contacting surfaces, z is the internal friction state and can be interpreted as average bristle deflection. F_f is the overall friction.

$$F_f = \sigma_0 z + \sigma_1 \frac{dz}{dt} + \sigma_2 v$$

$$\dot{z} = v - \sigma_0 \frac{z}{g(v)} |v|$$

$$g(v) = F_c + (F_s - F_c) e^{-(v/v_s)^\alpha} \tag{4-1}$$

The LuGre model generates a spring-like friction behaviour for small displacement, in which dominated by stiffness parameter σ_0 and micro-damping σ_1 . The $\sigma_2 v$ is viscous friction, this part can be replaced with a function $f(v)$ to represent macro-damping. The $g(v)$ captures Coulomb friction and Stribeck effect, the parameter v_s is Stribeck velocity and it determines how fast $g(v)$ approach F_c . Different scenarios have corresponding different parameter α .

The LuGre model above is a model dependent on velocity only, and most friction models are used to predict friction from mechanical motion transmission elements such as gears, screws and electrical motors. Friction in these applications is most metal-to-metal contacts, with different lubrication conditions. But in hydraulic cylinders, elastomeric seals have a distinct influence[79].

Pressure term is introduced in a friction model to interpret that the friction value is dependent on load[80], and when the load motion reverses at low velocity (nearly zero), a velocity function is not adequate to depict the friction[80]. A pressure-dependent friction model is applied in a hydraulic motor[81], which is capable of describing friction at low velocity. These examples revealed that pressure is a distinct factor that will affect friction in hydraulic applications.

The types of seals in the cylinder have different friction characteristics, the section view of the asymmetric cylinder utilised in this research is depicted in Fig. 4-1, it can be observed that its piston seal is a double acting H shape seal.

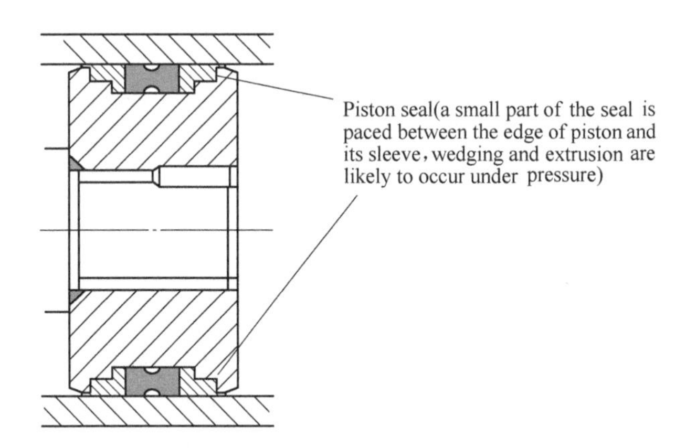

Fig. 4-1 C10 series 18 asymmetric cylinder piston seal[75]

The piston seal, the rod seal and wiper seal, as in Fig. 4-2, will also affect the overall friction value. The friction will be affected by multiple factors, for instance, seal design and material, fluid and fluid pressure, temperature, rubbing speed and surface finish[82].

As the friction causes heat during operation and leads to degradation of the seals, it will show influence on friction again. The basic seal friction equation is given as below, it must be noticed that the friction coefficient μ is an empirical factor and varies with speed, time, material, surface roughness, etc.

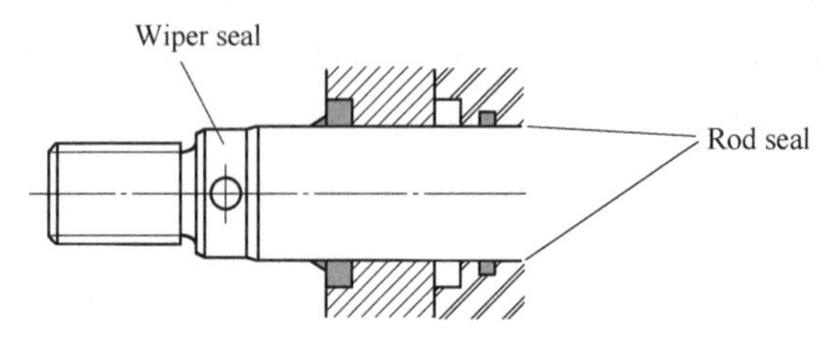

Fig. 4-2 C10 series 18 asymmetric cylinder rod seal and wiper seal[75]

The seal friction is proportional to the effective contact pressure[82], and can be expressed by an empirical equation as below:

$$\text{Seal friction} = \mu(v)p_e \cdot A \tag{4-2}$$

where μ is coefficient of friction; p_e is effective contact pressure; v is rubbing velocity; A is contact area.

In a normal situation the p_e effective pressure is equal to the pressure difference on both sides of the cylinder piston. The coefficient of friction μ is varied with speed v. The variable condition influence on the value of friction coefficient μ is listed in Table 4-1.

Table 4-1 Friction coefficient μ under various conditions[82]

Condition	Value
Dry seal surface	0.4 ~ 1
Lubricated elastomers surface	0.02 ~ 0.1
Lubricated fabric surface	0.04 ~ 0.08

The general seal friction Eq. (4-2) is feasible when the seal is not wedging, but the asymmetric cylinder piston seal in Fig. 4-1 is made by nitrile rubber and polyurethane. Its H shape structure makes it flexible and deformation under pressure, this design is to ensure the seal is always contacting with the surface.

This kind of seal called pressure-energised seals, is prone to extrusion and wedging under high pressures, especially when the clearance is generous (normally greater than 0.25mm)[82]. Under this circumstance, the friction force is increased considerably, and its value is proportional to the square of the effective pressure[82], so that the seal friction Eq. (4-2) becomes:

$$\text{Seal friction} = \delta \cdot \mu(v) \cdot (p_e)^2 A \tag{4-3}$$

where δ is a constant based on the seal type, m^2/N.

In this hybrid pump-controlled asymmetric cylinder system, a small part of the seal is placed between the edge of the piston and its sleeve as shown in Fig. 4-1, the design of the seal will tend to wedge and extrusion. Besides, this asymmetric cylinder is operated with long-term service, wears may occur in piston seals, which leads to generous clearance. But the rod seal and wiper seal are not under pressure during system operation, and they are equivalent to simple lubricated contacting surfaces. So that the friction of the rod seal and wiper seal should be depicted by velocity functions. The load coupling to the cylinder is moving horizontally on a plate, so its friction should also be interpreted by a velocity function.

The LuGre model is a well-performed friction model based on the velocity, and the energy-pressurised seal friction is not a complex function. So that a new friction model based on the LuGre model is implemented as below.

$$\begin{aligned} F &= \sigma_0 z + \sigma_1 \dot{z} + K(p_d)^2 v \\ \dot{z} &= v - \sigma_0 \frac{|v|}{g(v)} z \\ g(v) &= F_c + (F_s - F_c)e^{-|v/v_s|^\alpha} \end{aligned} \tag{4-4}$$

where K is a gain for pressure-energised seal friction Eq. (4-4), $m^3 \cdot s/N$, it's composed of the friction coefficient μ, constant δ and the contact area A. The contact pressure p_e is not measurable and it is greatly influenced by pressure difference on the seal[82], so that the pressure difference p_d is utilised in Eq. (4-4) and the gain K will compensate for the difference between the p_e and p_d. As the parameter K is only obtained by curve fitting method, this proposed friction model is an empirical model.

In low velocity situation, stick-slip and Stribeck effects are mainly described by $\sigma_0 z + \sigma_1 \dot{z}$. When the cylinder is in the normal operation state, the pressure-energised seal friction part $K(p_d)^2 v$ will show its distinct influence. This new friction model should be able to capture both micro and macro friction behaviours of hydraulic applications. There are several parameters needs to be identified to complete the model, these processes will be carried out in the next section.

4.2 A unique figure in the data fog

A new friction model is implemented in the last section, whose parameters vary for many elements, for instance, the contacting surface roughness, loads, lubrication conditions. There should exist a unique set of parameters for the new friction model for the test rig that was implemented in Chapter 3, the parameters are listed in Table 4-2.

This section depicts the procedures of identification of all its parameters.

Table 4-2 Parameters to be decided for the new friction model

Parameter	Annotation
σ_0 /N · m^{-1}	Stiffness
σ_1 /N · s · m^{-1}	Micro-damping
K/m^3 · s · N^{-1}	Gain for seal friction
F_c /N	Coulomb friction
F_s /N	Stiction friction
v_s /m · s^{-1}	Stribeck velocity
α	Curve constant

4.2.1 No added load tests

This section describes the procedures of identification of parameters when the system is in a steady state and dynamic state in the first part, the second part compares the original LuGre friction model performance for this test rig with the performance of the new friction model.

Several tests are run to identify these parameters, and the first step is to find the value of the parameters when the asymmetric cylinder is in a steady state. A biased square wave command is sent to the system to perform a symmetric velocity square wave motion. The pressure transducers record the chamber pressure p_1 and p_2 , and the potentiometer records the displacement of the cylinder in real time. Cylinder velocity is calculated by the differential of the displacement by Eq. (4-5).

$$v = \frac{\text{Displacement difference}}{t_{\text{interval}}} \tag{4-5}$$

where t_{interval} is the time interval of each recorded displacement data, s.

The friction force in the square waves tests is estimated based on the steady state equations of the new friction model. The velocity is a constant when the asymmetric cylinder in a steady state, the internal state z or average bristle deformation z will remain as a constant, hence, its derivative $\dot{z} = 0$.

Let

$$\dot{z} = 0 = v - \sigma_0 \frac{|v|}{g(v)} z \tag{4-6}$$

Cancel the velocity v, reveals:

$$g(v) = \sigma_0 z \tag{4-7}$$

Substitute the steady state $\sigma_0 z$ into friction force Eq. (4-4) reveals:

$$F_{fss} = F_c + (F_s - F_c)\, e^{-\left|\frac{v}{v_s}\right|^{\alpha}} + K(p_d)^2 v \tag{4-8}$$

So that Eq. (4-8) is suitable to calculate the friction force when the asymmetric cylinder operates in a square waveform, all the unknown parameters in it can be obtained by curve fitting method. The curving fitting tool in Matlab requires experimentally measured friction force, but it is not possible to be directly measured. However, it can be obtained by Eq. (4-9), as the cylinder is placed horizontally, the preload pressure is not in the consideration.

$$F_f = p_1 A_1 - p_2 A_2 - ma \tag{4-9}$$

where m is the mass attached to the cylinder, kg; a is the acceleration, m/s^2.

When the asymmetric cylinder in a square wave motion, the cylinder acceleration will drop to zero after the cylinder switches its direction. Therefore, the Eq. (4-9) becomes Eq. (4-10).

For a steady state test, the friction force is calculated by:

$$F_f = p_1 A_1 - p_2 A_2 \tag{4-10}$$

For example, a biased square wave command is sent to the asymmetric cylinder to perform ±25 mm/s 0.2 Hz square wave motion, the chamber pressure p_1 and p_2 are measured by transducers and the friction force can be calculated by Eq. (4-10), its steady state friction force from experimental measurements is depicted in Fig. 4-3.

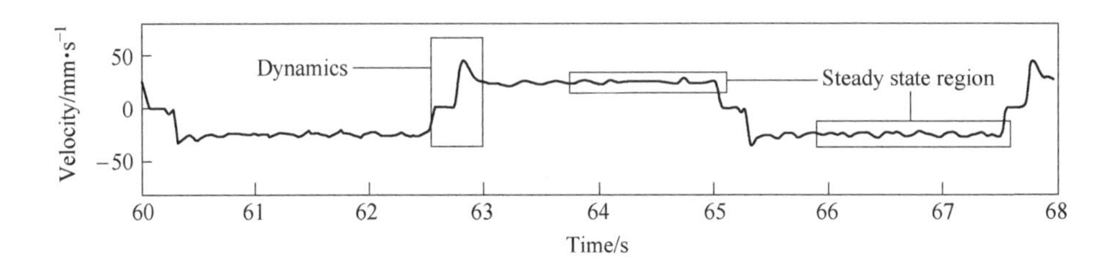

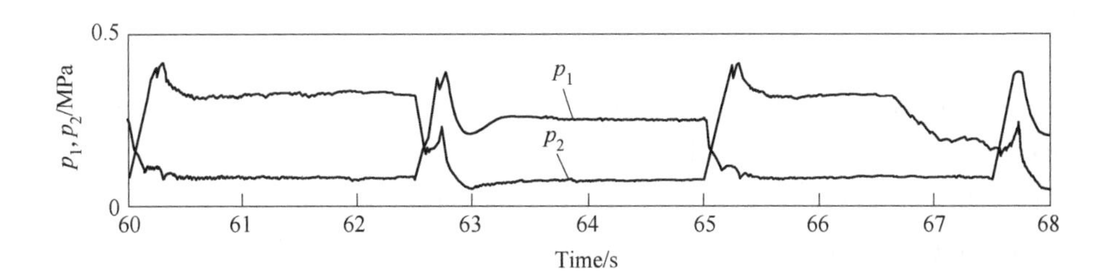

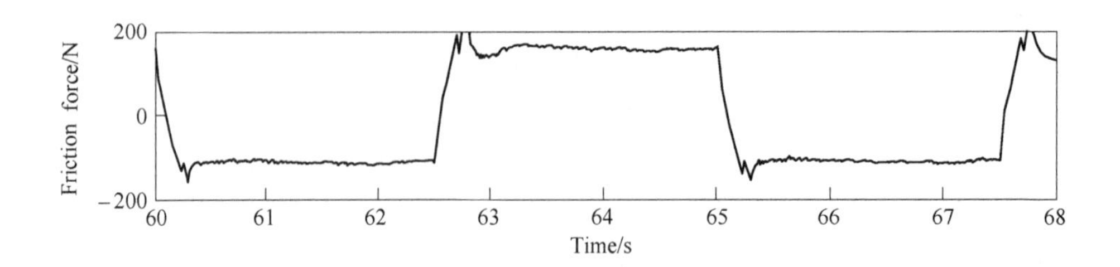

Fig. 4-3 Exampled calculated steady state friction force from ±25 mm/s measured data

The steady state friction force is obtained by averaging the friction values when the cylinder in steady state region as in Fig. 4-3, more square wave velocity inputs are sent to the test rig to measure the friction forces under different velocities. Its steady state friction forces and corresponding velocities are collected in Table 4-3.

Table 4-3 Steady state experimental measurements

Velocity/mm · s^{-1}	Friction force/N
48. 03	220
36. 18	181. 5
25	167
15	139. 2
0	180
0	150
0	-150
0	-188
-14. 73	-110
-24. 67	-119. 6
-36. 71	-136
-49. 05	-162. 2

There are some friction force values at zero velocity, which is due to the cylinder is in stiction state. The steps of no added load test can be summarised as:

(1) Disconnect the tray and cylinder.

(2) Send a biased square signal to the system to perform a symmetric square wave motion on the asymmetric cylinder.

(3) Record the peak (stiction friction) and steady state friction force.

(4) Repeat above processes with different velocity tests and draw the steady state friction force curve as in Fig. 4-4.

So, the steady state friction with different velocities is collected in Fig. 4-4.

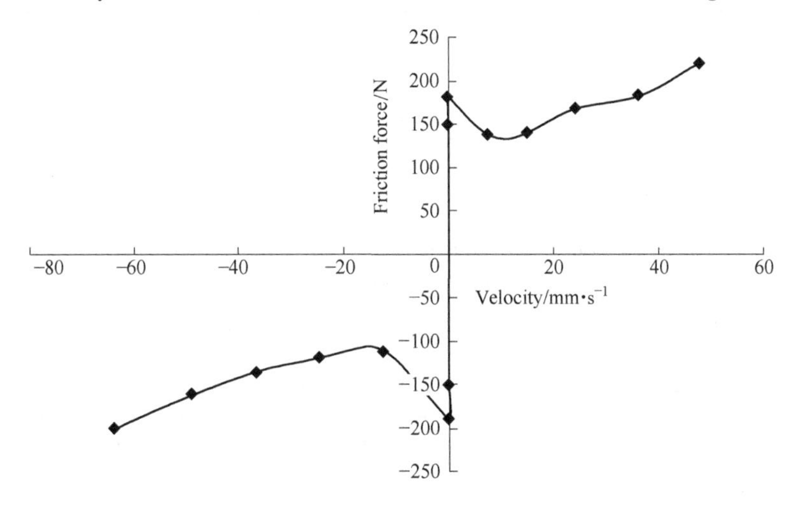

Fig. 4-4 Measured steady state friction force curve from various square wave tests

The right half of this curve is when cylinder in the extending mode, the left half is when cylinder in the retracting mode. In both modes, the stiction, Stribeck effect and viscous friction can be observed. When asymmetric cylinder starts its motion, the friction force reaches the maximum stiction friction force, then the friction force decreased with the increased velocity at low cylinder velocity. After the velocity exceeds some threshold, the friction force is increased with a larger velocity.

It can be noticed that when the asymmetric cylinder is operating at the same absolute velocity, the absolute friction force in the extending state is larger than that in the retracting state. This phenomenon is caused by the piston area difference. When the cylinder is retracting, less flow rate is required to achieve the same velocity in extending state, so the pressure in the rod side chamber is smaller. The difference can also be found in the pressure-energised piston seal, a smaller compress pressure will lead to a smaller friction force.

The effective pressure here is the pressure difference between both cylinder chamber, making the steady state friction F_{fss} Eq. (4-8) become:

$$F_{fss} = F_c + (F_s - F_c)\, e^{-\left|\frac{v}{v_s}\right|^{\alpha}} + K(p_1 - p_2)^2 v \tag{4-11}$$

Chamber pressures p_1, p_2 are obtained by pressure transducers, velocity is the derivative of measured displacement from the potentiometer. Observe the measured steady state friction force curve, some initial guessed parameters values are set as in Table 4-4.

Table 4-4 Initial guess value based on observation

Parameter	Extending	Retracting
F_c /N	110	-80
F_s /N	180	-188
v_s /m · s^{-1}	0. 01	-0. 01

The data collected in Table 4-3, the steady state friction force Eq. (4-5) and initial guess value in Table 4-4, put the data above into the curve fitting tools (CFTOOL) in Matlab, revealing the values of the parameters in Table 4-5.

Table 4-5 Parameter value obtained from CFTOOL in Matlab

Parameter	Extending	Retracting
α	0. 1087	0. 1138
K/m^3 · s · N^{-1}	2. 5×10^{-8}	4×10^{-9}

Substitute these values into the new friction model equations to obtain the simulated friction force. The measured friction curve and simulated data under a steady state are compared in Fig. 4-5, indicating that the new friction model is able to capture the friction behaviours in the hybrid pump-controlled asymmetric cylinder drive system when it is in a steady state.

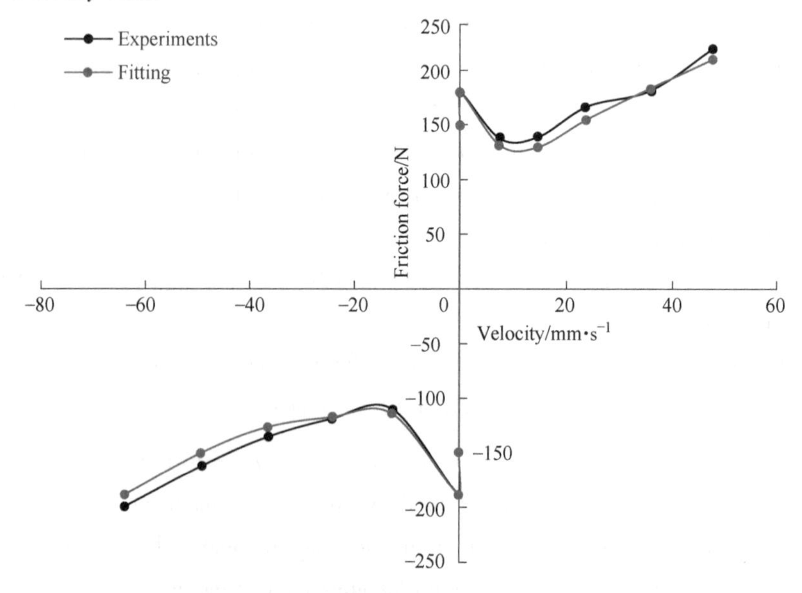

Fig. 4-5 Experimental data vs. simulated friction force results in steady state from square wave test

The steady state friction curve generated by the new friction model is able to capture the stiction, Stribeck effect and viscous friction phenomenon.

The σ_0 stiffness and σ_1 micro-damping is related to friction force dynamics, in the new friction model they combine with average bristle deformation z. The derivative of the average deformation $\dot{z}$ is not zero when the asymmetric cylinder keeps changing its velocity. To identify the value of σ_0 and σ_1, a sinusoidal wave of 0. 2 Hz command is sent to the system to make the asymmetric cylinder perform a sine wave motion with ±50 mm/s amplitude.

The measured dynamic friction force with 0. 2 Hz sinusoidal wave command is recorded in Fig. 4-6, where some spikes can be observed when the asymmetric cylinder changes its direction. The spikes are typically a stiction friction phenomenon combined with some cylinder dynamics, the pressure accumulates until the pressure difference is large enough to overcome the stiction force. When the piston starts to move, chamber pressure will suddenly drop to some level, the pressure drop also affects the friction force, shown as a spike in Fig. 4-6.

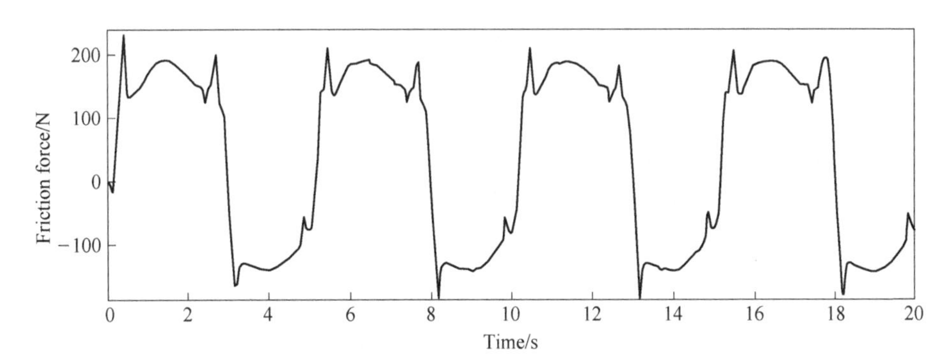

Fig. 4-6 Measured dynamic friction force on without added load (±50 mm/s sine wave)

With the known parameters, pick the guessed values for σ_0 and σ_1 as in Table 4-6, which are from a similar test rig in reference[83].

Table 4-6 Initial guesses of σ_0 and σ_1[83]

Parameter	Value
σ_0 /N · m^{-1}	1×10^7
σ_1 /N · s · m^{-1}	0. 1

The value of bristle stiffness σ_0 affects the magnitudes and rise time of the break-away force[84], and adjust its value until the simulated curve matches the experimental results. So, the values are adjusted as in Table 4-7.

Table 4-7 Adjusted values of σ_0 and σ_1

Parameter	Value
σ_0 /N · m^{-1}	3×10^7
σ_1 /N · s · m^{-1}	0. 1

Compare the experimental results with the new friction model as in Fig. 4-7.

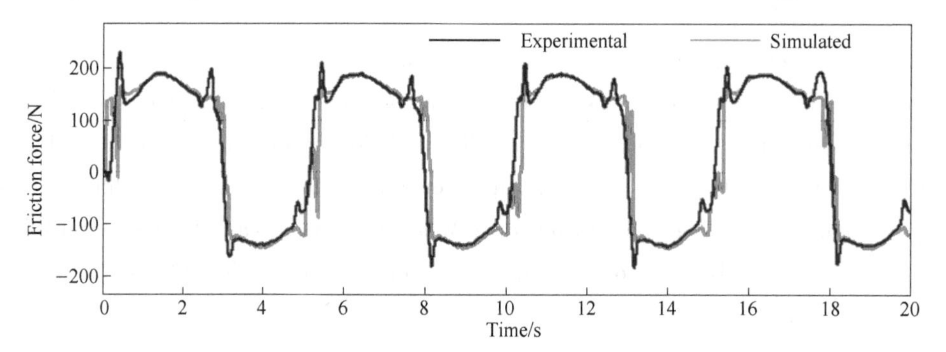

Fig. 4-7 No added load experimental dynamic friction vs. new friction model simulation results with pressure difference term (±50 mm/s sine wave)

Overall, the new friction model almost captures the dynamics in experimental, but the simulated friction is not catch up with the force when cylinder changes its direction, a clear view of friction force vs. velocity reveals this concern in Fig. 4-8.

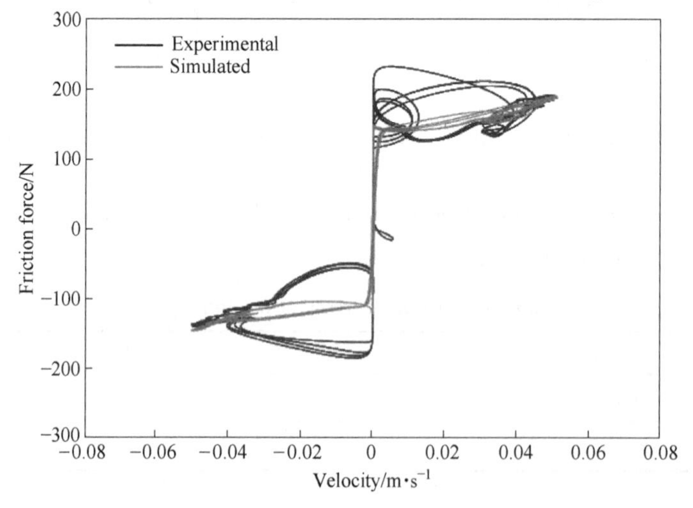

Fig. 4-8 No added load experimental friction compares with simulated friction with pressure difference term (±50 mm/s sine wave)

This phenomenon is related to the fluid film thickness changes during the piston acceleration and deacceleration[82], which means there is an acceleration term missing in the new friction model. To compensate the difference, the new friction model is updated as:

$$F = \sigma_0 z + \sigma_1 \dot{z} + K[(p_1 - p_2)^2 v + H \cdot a]$$
$$\dot{z} = v - \sigma_0 \frac{|v|}{g(v)} z$$
$$g(v) = F_c + (F_s - F_c)\, e^{-|v/v_s|^{\alpha}} \tag{4-12}$$

The H in Eq. (4-12) can be regarded as a gain ($N^2 \cdot s/m^4$), the CFTOOL in Matlab reveals that when the cylinder is extending, the $H = 3.046 \times 10^9$, while in retracting state, the $H = 9 \times 10^9$. Substitute the H back to new friction model, and compare with experimental results as in Fig. 4-9 and Fig. 4-10.

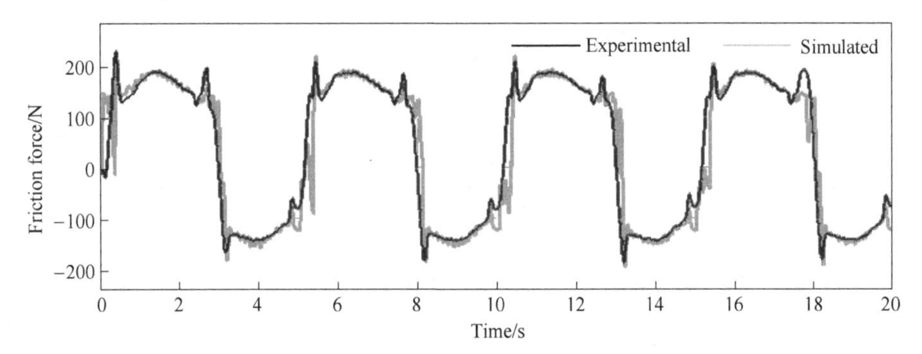

Fig. 4-9 No load experimental friction results vs. new friction model results with pressure and acceleration term (±50 mm/s sine wave)

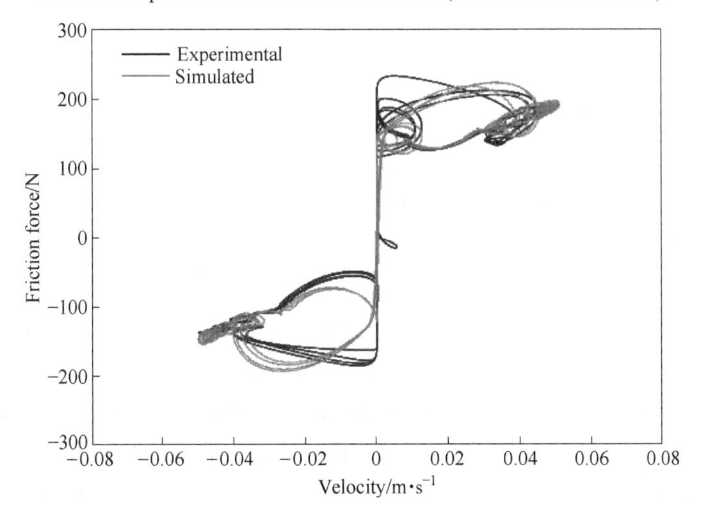

Fig. 4-10 No load experimental vs. simulation results of force vs. velocity with pressure and acceleration term (±50 mm/s sine wave)

The updated friction model with the acceleration term captures the dynamics of friction better during system operation. For this hybrid pump-controlled asymmetric cylinder drive system in this research, its new friction model parameters are listed in Table 4-8.

Table 4-8 Parameters of the new friction model for the test rig

Parameter	Extending	Retracting
F_c /N	110	-80
F_s /N	180	-188
v_s /m · s^{-1}	0. 01	-0. 01
α	0. 1087	0. 1138
K/m^3 · s · N^{-1}	2. 5×10^{-8}	4×10^{-9}
σ_0 /N · m^{-1}	3×10^7	3×10^7
σ_1 /N · s · m^{-1}	0. 1	0. 1
H/N^2 · s · m^{-4}	3. 05 × 10^9	9 × 10^9

From the results in Fig. 4-9 and Fig. 4-10, though the new friction model is not able to capture all the friction results from experimental perfectly, some transient changes are not followed, but it is almost consistent with experimental results. So, the new friction model can be judged as a success. These parameters in Table 4-8 are obtained without a load applied to the cylinder, the next section discusses the load influence on the parameters.

4. 2. 2 Added load tests

All the above tests are carried out without any load, in theory, the load should only affect stiction friction F_s and Coulomb friction F_c [84]. Two loaded tests are carried out to verify this assumption, the first test is a total 25 kg load (tray and added load) attached to the asymmetric cylinder. Same biased command of 0. 2 Hz sinusoidal wave is sent to the cylinder to perform the same sine wave motion as no added load test.

Keeps the values of parameters the same as no load except the stiction friction F_s and Coulomb friction F_c , adjust their value as in Table 4-9, simulate the new friction model with Eq. (4-12) and compare with the measured friction force as in Fig. 4-11.

Table 4-9 Stiction force F_s and Coulomb force F_c for 25 kg load

Parameter	Extending	Retracting
F_c /N	120	-90
F_s /N	200	-190

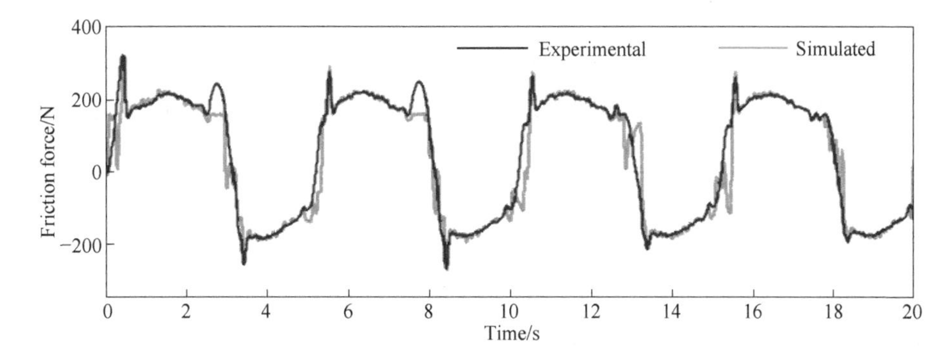

Fig. 4-11 Added 25 kg load friction force of experimental & new friction model simulation results (±50 mm/s sine wave)

Observing these results in friction force vs velocity form in Fig. 4-12, the new friction model is still able to capture almost all the experimental friction behaviours under the loaded condition.

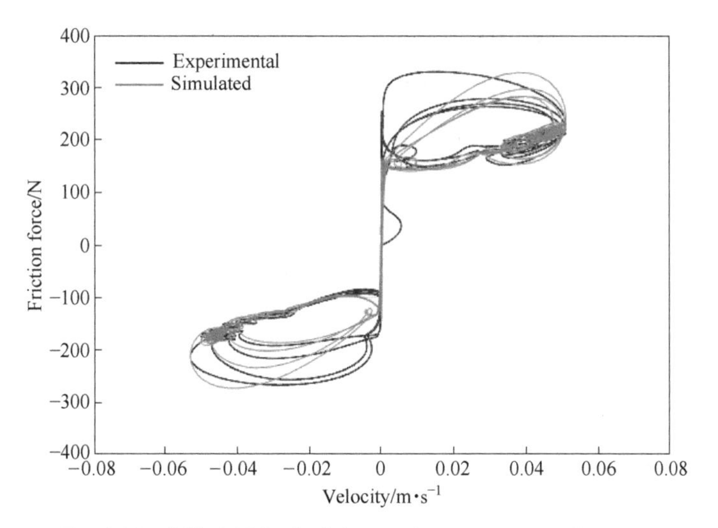

Fig. 4-12 Added 25 kg load friction force experimental vs. new friction model simulation results (±50 mm/s sine wave)

A further test is carried with a 38 kg load (tray and added load) attached to the asymmetric cylinder, changing the stiction friction F_s and Coulomb friction F_c as in Table 4-10.

Table 4-10 Stiction force F_s and Coulomb force F_c for 38 kg load

Parameter	Extending	Retracting
F_c /N	122	-80
F_s /N	205	-188

Simulate the friction force and compare with the experimental results as in Fig. 4-13.

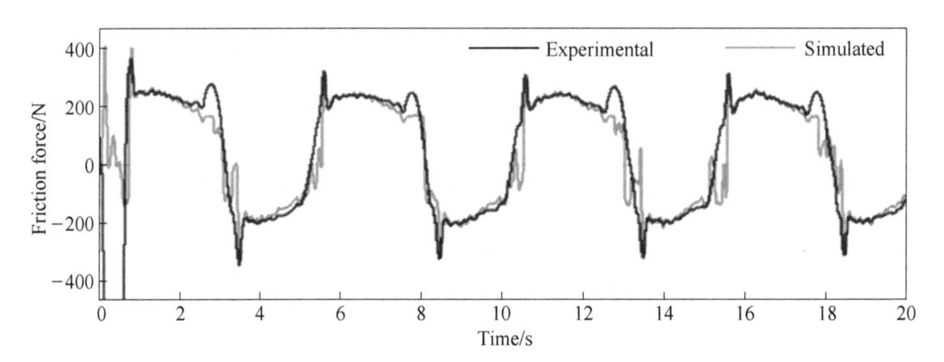

Fig. 4-13 Added 38 kg loaded friction force experimental & new friction model simulation results (±50 mm/s sine wave)

The simulated curve shows good consistency with experimental results.

Observe the friction force vs. velocity curves in Fig. 4-14, though some transient

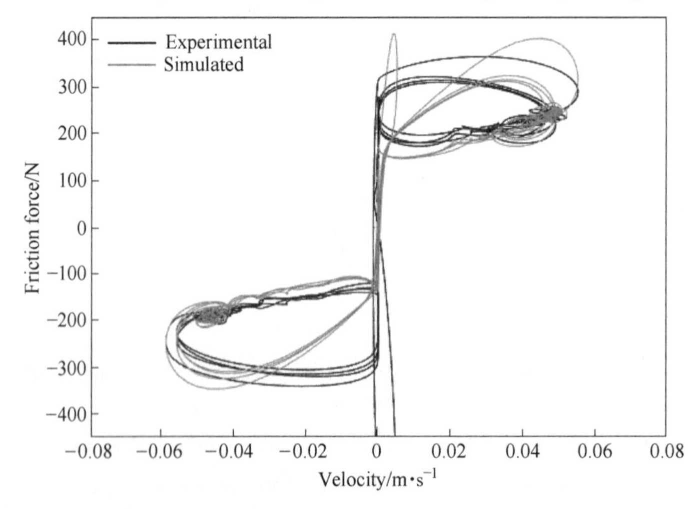

Fig. 4-14 Added 38 kg loaded friction force experimental vs. new friction model simulation results (±50 mm/s sine wave)

behaviours are not captured, the new friction model still captures almost all the friction behaviours in the hybrid pump-controlled asymmetric cylinder drive system. So that the assumption has been verified, the external load only affects the parameter F_c and F_s. However, the values of both parameters are not obviously changed with different added load, this may due to the added load is already reflected on the pressure term of the new friction model. All these tests are carried out in a low velocity scenario, the performance of this new friction model under high velocity will be tested in future.

4.3 Contrast shows the truth

The performance of the new friction model is verified in the previous section, but how much the new friction model is improved compared to the original LuGre model is discussed in this section. The original LuGre friction model can be expressed by:

$$F_f = \sigma_0 z + \sigma_1 \frac{dz}{dt} + \sigma_2 v$$

$$\frac{dz}{dt} = v - \frac{z}{g(v)}|v|$$

$$g(v) = F_c + (F_s - F_c)\, e^{-(v/v_s)^{\alpha}}$$

The values of parameters are set as in Table 4-11.

Table 4-11 LuGre friction model parameters for no-load test

Parameter	Extending	Retracting
F_c /N	110	-80
F_s /N	180	-188
v_s /m · s^{-1}	0.01	-0.01
α	0.1087	-1138
σ_0 /N · m^{-1}	3×10^7	3×10^7
σ_1 /N · s · m^{-1}	0.1	0.1
σ_2 /N · s · m^{-1}	1000	900

The LuGre friction model is utilised in no-load set up, the set up is to minimise the influence from the external environment. Run the simulation and compare to the experimental results as in Fig. 4-15.

No matter what values are given to its parameters, the simulated LuGre friction model is not able to capture the experimental measurements as good as the new friction model results in Fig. 4-9.

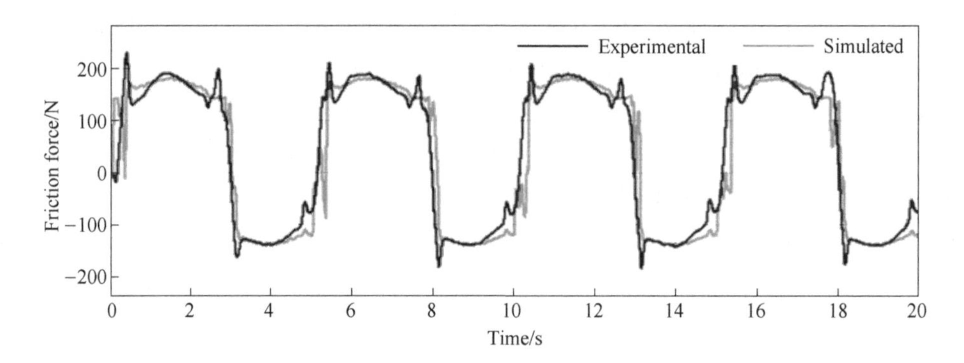

Fig. 4-15 Original LuGre model friction force compared with experimental results

When observing the friction force vs. velocity results in Fig. 4-16, the simulated LuGre friction model misses a lot of friction dynamics in the hybrid pump-controlled asymmetric cylinder drive system.

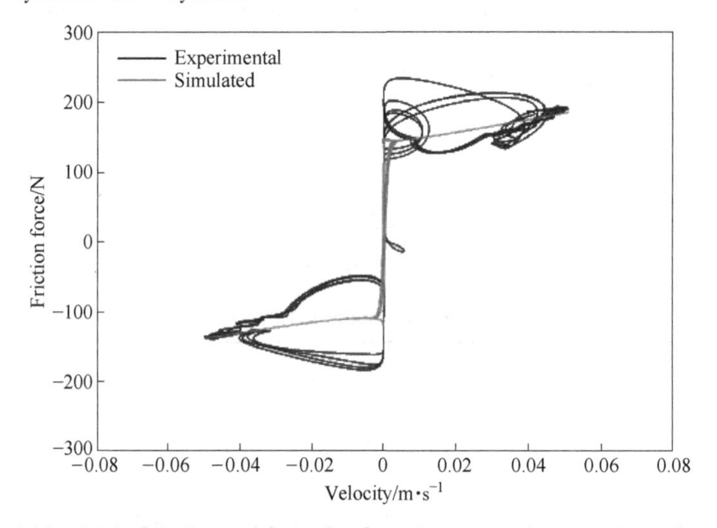

Fig. 4-16 Original LuGre model simulated results compared to experimental results

So that for the hydraulic applications, the new friction model is an improved solution to simulate friction force.

4.4 Concluding remarks

This chapter implements a new friction model developed on the LuGre model, the original LuGre model is described below.

$$\text{Orignal LuGre model}\begin{cases} F_f = \sigma_0 z + \sigma_1 \dfrac{dz}{dt} + \sigma_2 v \\ \dot{z} = v - \sigma_0 \dfrac{z}{g(v)} |v| \\ g(v) = F_c + (F_s - F_c) e^{-(v/v_s)^\alpha} \end{cases} \tag{4-13}$$

The new friction model combines friction calculation with velocity, chamber pressure and acceleration, developed a new friction model that can be applied for hydraulics application as below:

$$\text{New friction model}\begin{cases} F_f = \sigma_0 z + \sigma_1 \dot{z} + K[(p_1 - p_2)^2 v + H \cdot a] \\ \dot{z} = v - \sigma_0 \dfrac{|v|}{g(v)} z \\ g(v) = F_c + (F_s - F_c) e^{-|v/v_s|^\alpha} \end{cases} \tag{4-14}$$

Compared with the original LuGre model, some new parameters are introduced, the $K[(p_1 - p_2)^2 v]$ part added the pressure influence in the $\sigma_2 v$ part in original LuGre model, and an extra $KH \cdot a$ part describes the influence brought by the acceleration. As the parameter K and H are obtained by curve fitting method, the new friction model is an empirical model. Their performances are compared in Fig. 4-17.

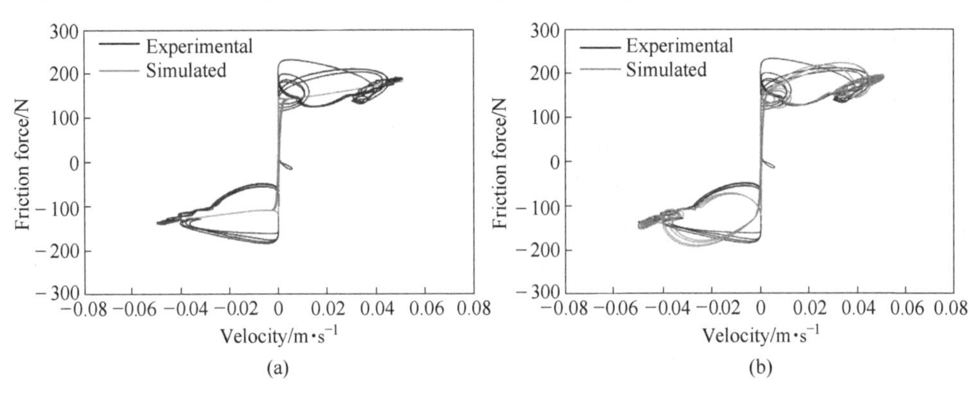

Fig. 4-17 Original LuGre friction model performance vs. new friction model performance

(a) Original LuGre model performance; (b) New friction model performance

Its simulation friction results are compared with experimental results when asymmetric cylinder performs ±50 mm/s sine wave motion, validated its feasibility to describe the friction force in a hybrid pump-controlled asymmetric cylinder drive system. Compared with the simulation results from the original LuGre friction model, the improvements of the new friction model are validated. This new friction model will be added to the simulation model for the hybrid pump-controlled asymmetric cylinder drive system for a better result.

Chapter 5 System Modeling

This chapter utilises the component linking method from Chapter 2 to simulate the open loop tests of the hybrid pump-controlled asymmetric cylinder drive system in Chapter 3, and aims at the construction of a model to describe the test rig for nonlinearities analysis and further research. There is no mechanism to keep a constant supply pressure in the hybrid pump-controlled system, and the flow and force balance equations that utilised in the valve-controlled system are no longer suitable, but the algorithm of component linking is still applied.

5.1 Learn the new from the old

The hybrid pump-controlled asymmetric cylinder drive system in Fig. 5-1 is modeled to simulate its performance in velocity, force, pressure, etc. The power drive controller sends command signals to the motor, the coupling pump delivers the flow into the pipelines, the pilot four-way valve is shifted to the corresponding position and the fluid flow is delivered to cylinder chamber to drive the cylinder. The cylinder is coupling with a load and its motion is operated to follow the command signal.

Similar to the component linking model of the valve-controlled asymmetric drive system in Chapter 2, this open loop system model is composed of a pump model, valve model, actuator model and load model. The major difference is that the command was sent to the valve to adjust its valve opening to control the chamber pressures in the valve-controlled system, while this hybrid pump-controlled system accepts the signal by the pump and the pump delivers flow into service lines to create chamber pressures to drive the asymmetric cylinder. Its algorithm is depicted as in Fig. 5-2, each component is modeled individually and combined to simulate the whole system.

This hybrid pump-controlled asymmetric drive system is a partially closed circuit. When the cylinder is extending, all the fluid in the extending service line will be delivered to the piston side chamber, and all the fluid in the retracting line will go back

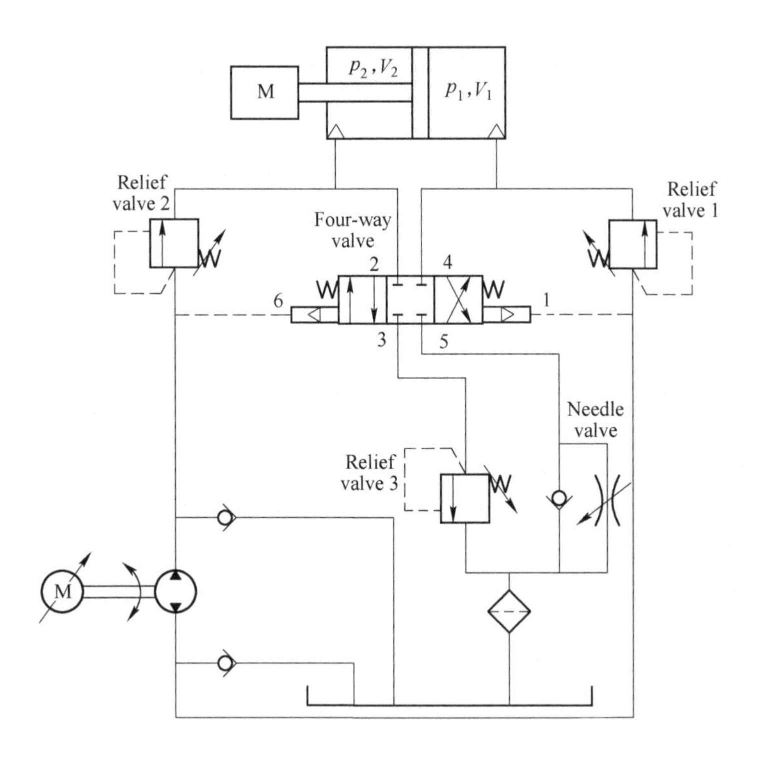

Fig. 5-1 Hybrid pump controlled asymmetric cylinder drive system

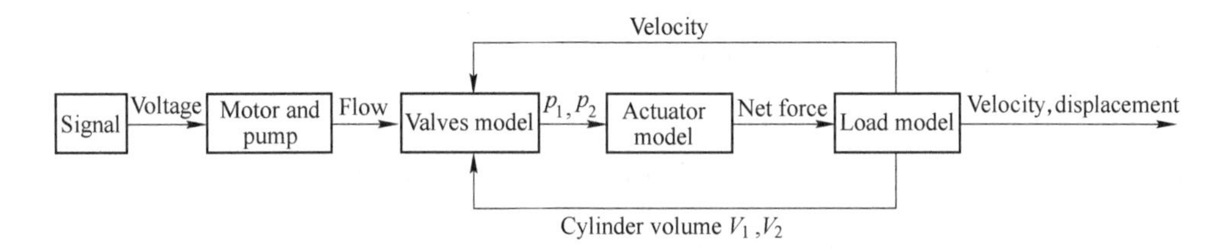

Fig. 5-2 The schematic of modeling the hybrid pump controlled asymmetric cylinder drive system

to the tank like a valve-controlled system. The pilot four-way valve is oversizing to reduce the throttle losses and the pump delivers the flow to the cylinder chamber to generate required pressures, this design targets at reducing the major energy losses in the valve-controlled system. The needle valve in the return line is to increase the stability during operation but will introduce some throttle losses. So that this proposed design is to find a balanced middle place between a valve-controlled and a pump-controlled asymmetric cylinder drive system, the rest of this chapter will build a system model to verify and investigate its performance.

5.2 Learn mathematically about component models

This section simulates the major hydraulic components of the hybrid pump-controlled asymmetric cylinder drive system, starting with the motor and pump, then valves, and the asymmetric cylinder drive. Utilise component linking algorithm to model the whole system.

5.2.1 Modeling of the motor and pump

The bidirectional HPI gear pump is coupling with the servomotor, the motor drive sends command voltages to the servomotor, then the servomotor sends speed information from tachometer to the amplifier drive to ensure the servomotor's speed follows the reference command. The coupling gear pump operates at a corresponding velocity to deliver the fluid flow into the service line at a certain rate. The process is described as in Fig. 5-3, it can be regarded as a command is input to a block and output flow rates to the system.

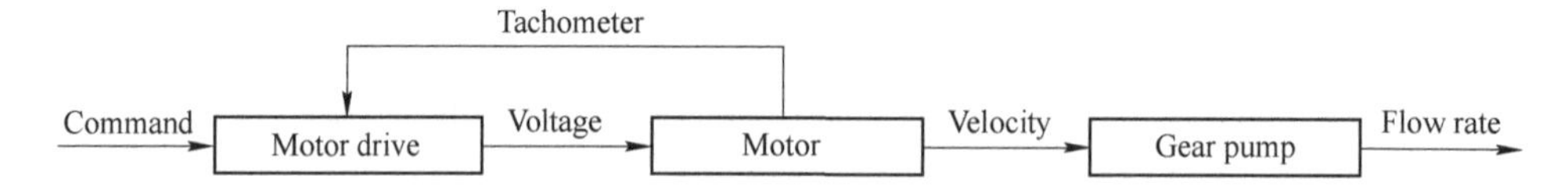

Fig. 5-3 Pump and motor modeling scheme

Using the parameter of gear pump capacity data (from Chapter 3) to covert the motor velocity to flow rate, the scheme in Fig. 5-3 can be simplified as in Fig. 5-4.

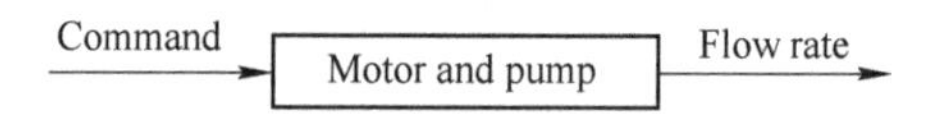

Fig. 5-4 Simplified pump and motor model

The motor and pump model can be viewed as a second order transfer function with a gain, natural frequency and damping, the gain is described as below:

$$\text{Signal(V)} \times 1.07 \times 10^{-5} = \text{Flow rate}(\text{m}^3/\text{s})$$

In Matlab Simulink, the pump and motor model are depicted in Fig. 5-5.

The transfer function in Fig. 5-5 shows that the natural frequency is 47.62 rad/s and the damping ratio is 0.64. Both values are identified by curve fitting tools, and the experimental step response of motor is depicted in Fig. 5-6.

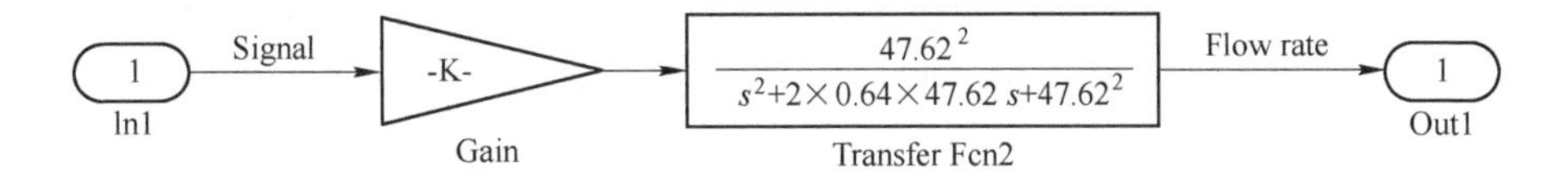

Fig. 5-5 Pump and motor model in Simulink

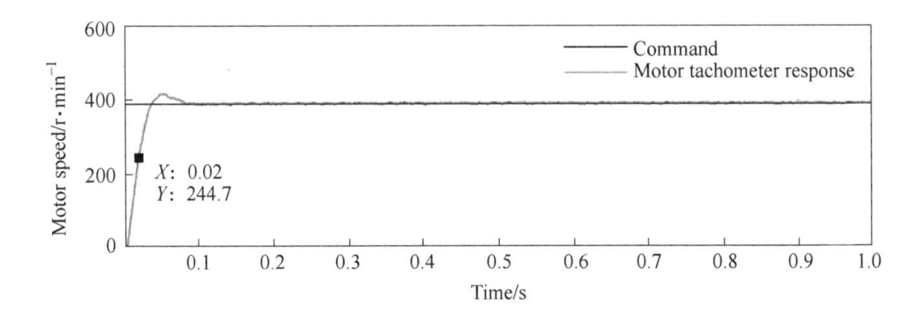

Fig. 5-6 Step command and motor tachometer response

5.2.2 Valves model

There are three types of valve in the hybrid pump-controlled asymmetric cylinder drive system, a pilot four-way directional valve, three pressure relief valves and a needle valve. This section will simulate these three types of valve.

5.2.2.1 Pilot directional four-way valve

This pilot directional four-way valve is from SunHydraulics with a model number of DDDCXCN. A spring-centred, normally closed, piloted by the pressure difference between the port 1 and 6 four-way valve circuit is depicted in Fig. 5-7.

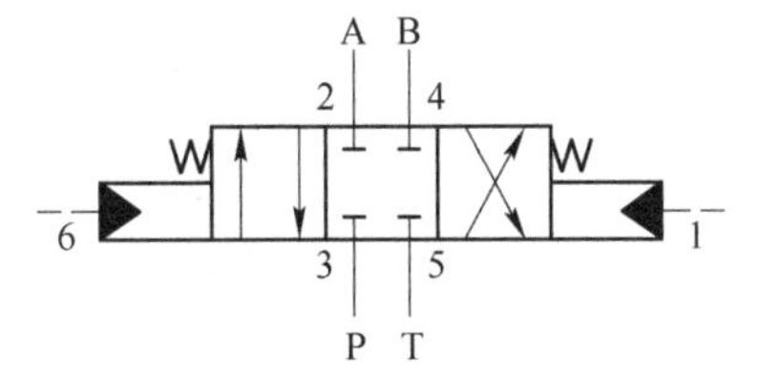

Fig. 5-7 SunHydraulic DDDCXCN model pilot directional four-way valve circuit

In the component linking simulation work done by Leaney[63], the supply pressure is constant and the cylinder chamber pressures are adjusted by the valve opening area. But in this hybrid pump controlled asymmetric cylinder drive system, there is no constant supply pressure p_s, and the valve porting details are not offered as well, which indicates that the pressure equations in Chapters 1 and 2 are not feasible for this application.

However, the performance curves in Fig. 5-8 from its datasheet[72] can be utilised to model this valve. The flow rate passing through a valve can be calculated by a general equation as below:

$$Q = C_D A_v \sqrt{\frac{2}{\rho}} \sqrt{\Delta p} \tag{5-1}$$

where Δp is the pressure difference between the valve inlet and outlet ports, N/m^2.

The flow vs. pressure performance curves in Fig. 5-8 are generated when the valve is shifted (the valve spool is placed at the left and right end), which indicates the valve opening A_v in the performance curves remain maximum.

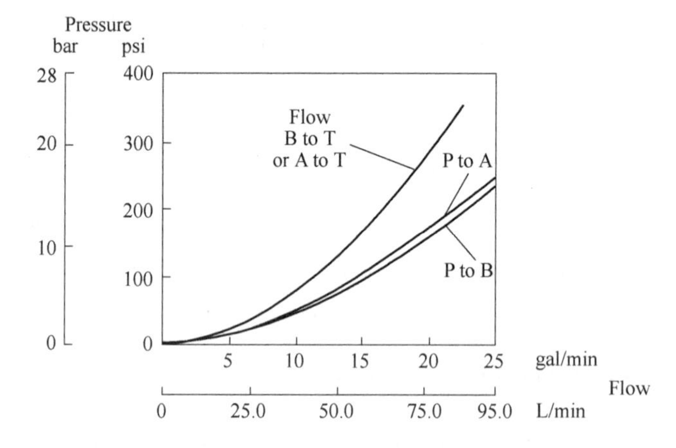

Fig. 5-8 DDDCXCN four-way valve performance curve[72]
(1 bar = 14.5 psi = 0.1 MPa)

The mechanism of this pilot-shifted four-way valve in this hybrid pump-controlled system should only operate at three positions, spring centred, left end and right end. This four-way valve is an oversizing valve, when the valve is maximum opened its capacity is much larger than the requirement of the system, so that the throttle losses can be reduced to the minimum level.

As the coefficient of discharge C_D, valve opening A_v and oil density ρ are constant, the valve performance in the system should be consistent with its performance curve. The $C_D A_v \sqrt{\frac{2}{\rho}}$ part in Eq. (5-1) can be viewed as a constant, it can be noticed that the curve of the B to T and A to T are placed higher than that of P to A and P to B, this difference is due to the porting designs of this valve is not all the same. To model the performance curve in an easier way, Let:

$$C_D A_v \sqrt{\frac{2}{\rho}} = K_T \text{ when T port connected with A and B} \tag{5-2}$$

$$C_D A_v \sqrt{\frac{2}{\rho}} = K_P \text{ when P port connected with A and B} \tag{5-3}$$

Selects points from both curves and calculate their average values reveals:

$$K_P = 1.0644 \times 10^{-6}\ \mathrm{m^4/(N^{\frac{1}{2}} \cdot s)}$$

$$K_T = 1.3730 \times 10^{-6}\ \mathrm{m^4/(N^{\frac{1}{2}} \cdot s)}$$

Substitute K_P, K_T and a set of values of Δp to model the performance curve

$$Q_P = K_P \sqrt{\Delta p} \text{ for port P} \tag{5-4}$$

$$Q_T = K_T \sqrt{\Delta p} \text{ for port T} \tag{5-5}$$

The simulated curves are revealed as in Fig. 5-9, which shows consistency with the performance curves from datasheet in Fig. 5-8.

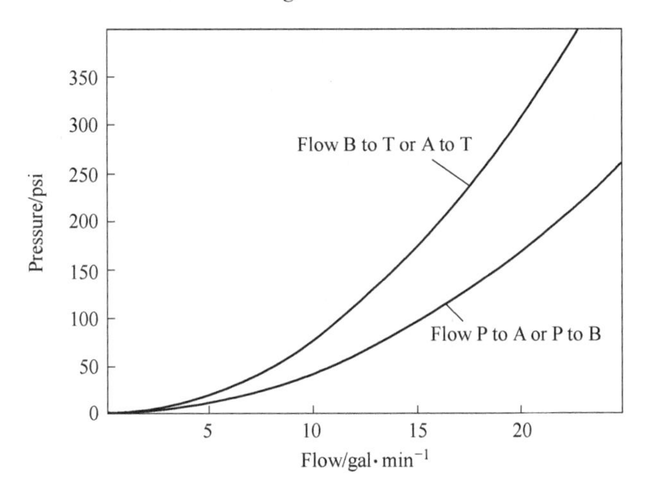

Fig. 5-9 Simulated DDDC four-way valve performance curves
(1 psi = 6.89 kPa, 1 gal = 4.55 L)

As the flow and pressure equations for this four-way valve are obtained, the algorithm of this pilot-shifted four-way valve is depicted in Fig. 5-10, the spool stroke x_s here is only used when the valve opening area is smaller than that of the needle valve, the details will be depicted later in section 5. 2. 2. 3.

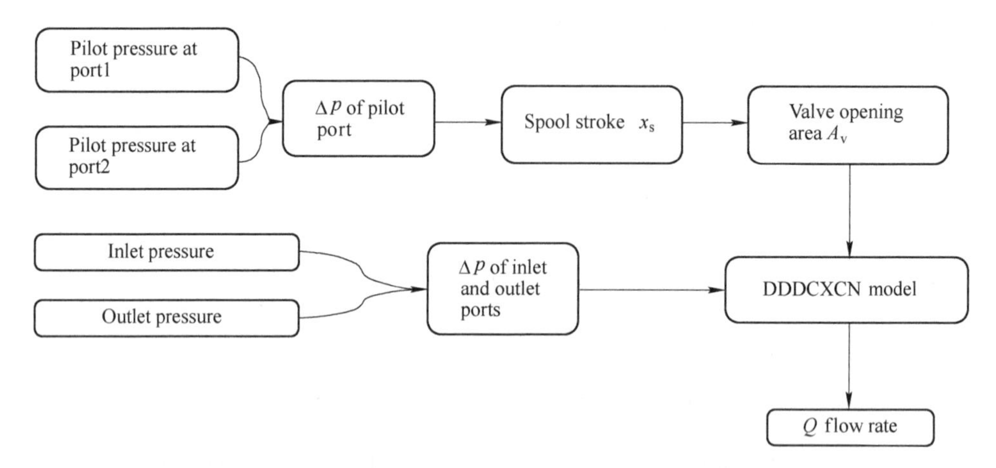

Fig. 5-10 DDDCXCN four-way valve model block diagram

5. 2. 2. 2 Direct-acting relief valve

This section is to model the RBAELAN model pressure relief valve from SunHydraulics, the relief valve is a two-port, pilot stage, normally closed pressure regulating valve with fully adjustable pre-settings. When the pressure at inlet port is adequate to overcome the pre-set spring force, a flow path is created from port 1 to port 2 as in Fig. 5-11.

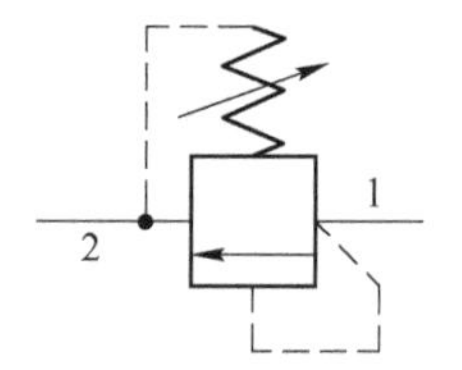

Fig. 5-11 SunHydraulics RBAELAN model valve symbolic scheme

Its function is basically the same as a conventional relief valve to regulate the pressure in the system. However, the RBAELAN model relief valve must consider the pressure at the outlet port, the fluid pressure at port 2 is directly added to the pressure setting at port 1. In a conventional valve system set up, the pressure relief valve's outlet port is connected to a tank or reservoir, their pressure is normally equal to zero (gauge pressure). Its function can be understood as the pressure at inlet port must overcome the pre-set spring force to open the valve. Two of this RBAELAN valve are placed in the service line, whose outlet port is connected to cylinder chamber inlet. But the pressure in the cylinder chamber is usually not equal to zero when the cylinder piston is moving during system operation. Thus, the pressure at port 2 must be taken into consideration.

The spring force can be adjusted in a range of 0. 17 MPa to 21 MPa, and its factory default setting is 7 MPa. The RBAELAN valve will perform differently under various pressure settings, and these performance curves are given in Fig. 5-12.

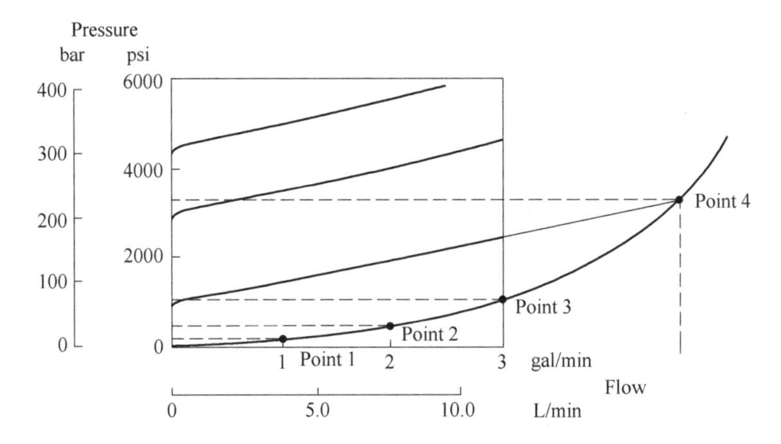

Fig. 5-12 RBAELAN model relief valve performance curves
(1 bar = 14.5 psi = 0.1 MPa)

The different performance curves in Fig. 5-12 indicate the relationship between flow and pressure difference with the corresponding pressure pre-settings of the relief valve. The datasheet[73] shows it is a poppet type valve, but the key parameters of the structure are not offered. In order to model this relief valve, a similar methodology of the four-way valve model in section 5.2.2.1 is utilised here, which uses its performance curve to model this relief valve.

The algorithm is inspired by a scheme curve that is shown in Fig. 5-13[85], when the inlet pressure does not reach the cracking pressure, zero flow rate is permitted. When the inlet pressure is higher than the cracking pressure, the opening area keeps increasing until the pressure difference increased by Δp_r. During this process, the flow

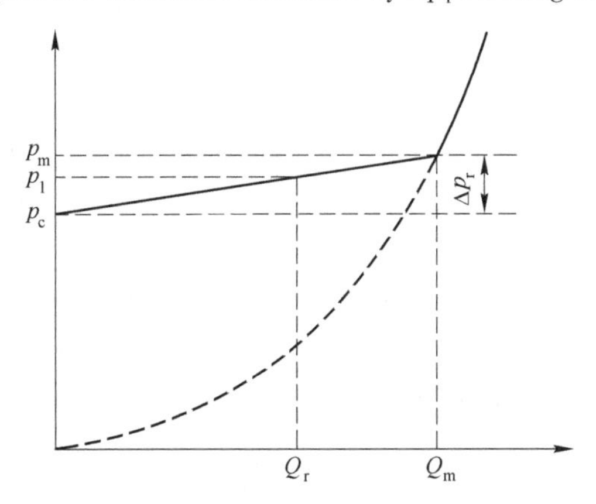

Fig. 5-13 Typical pressure relief valve performance curve[85]

rate can be expressed with a linear equation. By then, the valve reaches its maximum opening area, the flow rate is expressed by another equation after the inlet pressure is higher than p_m. These equations are described below:

$$Q_r = \begin{cases} 0 & \text{if } p_{In} < p_c \\ Q_m \dfrac{p_{In} - p_c}{p_m - p_c} & \text{if } p_c < p_{In} \leqslant p_m \\ K_v \sqrt{p_{In} - p_{Out}} & \text{if } p_{In} > p_m \end{cases} \tag{5-6}$$

where Q_r is the flow rate pass through the relief valve, m^3/s; Q_m is the flow rate when the valve reaches its maximum opening area, m^3/s; p_c is the pre-setting crack pressure, N/m^2; p_{In} here is stands for the inlet pressure, N/m^2; p_{Out} is the outlet pressure, N/m^2; K_v is the valve pressure-flow coefficient as:

$$K_v = \frac{Q_m}{\sqrt{p_m}} \tag{5-7}$$

To build this model, the Q_m and p_m are required, they are constants with a pre-set cracking pressure. These two parameters are the intersection point of two curves in Fig. 5-13, it can be solved by Eq. (5-8):

$$p_m^2 - (2p_c + R_a^2 K_v^2) p_m + p_m^2 = 0 \tag{5-8}$$

where R_a is the ratio of pressure and flow before the relief valve reaches maximum spool stroke, and it is used to model a part of the performance curve as:

$$R_a Q_r + p_c = p_r \tag{5-9}$$

After solving the Eq. (5-8), the value of p_m is revealed, by then, the Q_m can be calculated with parameter K_v. The value of K_v is obtained by measuring the pressure and flow rate of point 4 in Fig. 5-12, then utilise the Eq. (5-8) to obtain the value of K_v.

Use the Eq. (5-6) to draw performance curves with various cracking pressures and compare with the factory performance curves as in Fig. 5-14.

The comparison in the above Fig. 5-14 indicates that the Eq. (5-6) are able to mathematically describe the RBAELAN valve's flow and pressure behaviour. However, during the operational condition of the hybrid pump-controlled system, the outlet of the relief valve is not constantly zero pressure. Due to its function that the outlet pressure will directly be added to the inlet pressure, the Eq. (5-6) should be rewritten as:

$$Q_r = \begin{cases} 0 & \text{if } p_{In} - p_{Out} < p_c \\ Q_m \dfrac{p_{In} - p_{Out} - p_c}{p_m - p_c} & \text{if } p_c < p_{In} - p_{Out} \leqslant p_m \\ K_v \sqrt{p_{In} - p_{Out}} & \text{if } p_{In} - p_{Out} > p_m \end{cases} \tag{5-10}$$

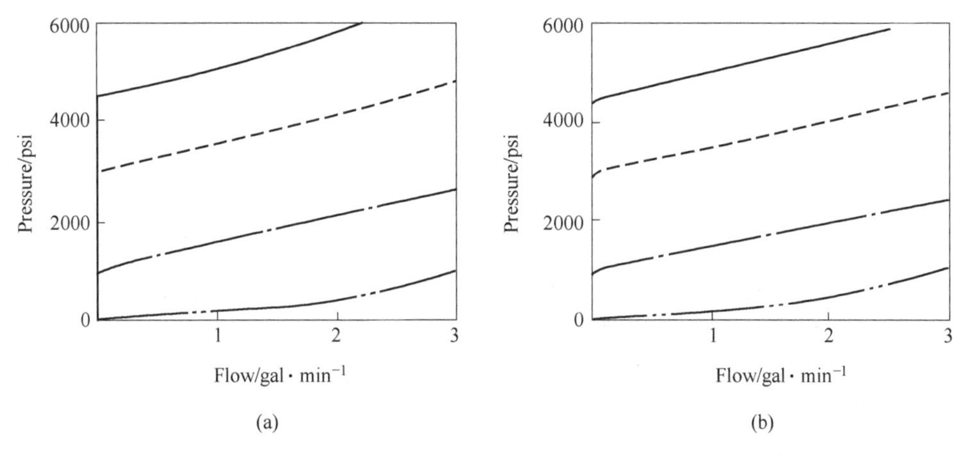

Fig. 5-14 Modelled performance curves (a) vs. factory data (b)
(1 psi = 6.895 kPa, 1 gal = 4.55 L)

5.2.2.3 Fully adjustable needle valve

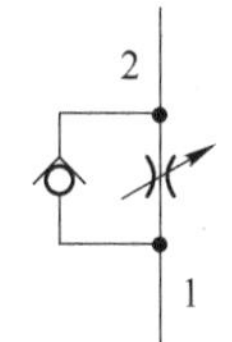

Fig. 5-15 NCCC needle valve diagram circuit[74]

This section models the NCCC needle valve from SunHydraulics, which is a needle valve with a reverse-flow check and fully adjustable function. Its diagram circuit is depicted in Fig. 5-15, this needle valve can be fully adjusted by turning the screw on top to regulate the flow passing through.

This valve is used to limit the flow at the return line of the hybrid pump controlled asymmetric cylinder by reducing its valve opening, the directional four-way valve is connected with the needle valve at the return line. Moreover, they are connected by a short stainless steel pipeline, the DDDC four-way valve is an oversizing valve compared to this needle valve, so that it can be regarded that the return line valve opening is only dominated by this needle valve.

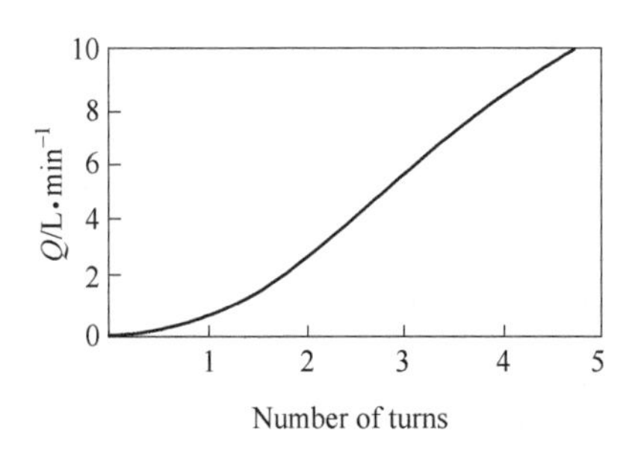

Fig. 5-16 NCCC needle valve performance curve[74]
Adjustment Sensitivity at 7 bar Differential

To model this simple function valve, its valve opening area is the only parameter that required. It still based on the performance curve in Fig. 5-16.

Utilise the general flow and pressure Eq. (5-1) and rearrange it into below Eq. (5-11), so that the valve opening area can be calculated corresponding to different screw turns.

$$A_{vn} = \frac{Q}{C_D \sqrt{\frac{2}{\rho}} \sqrt{\Delta p}} \tag{5-11}$$

where the coefficient of discharge C_D , oil density ρ and pressure difference Δp are constant in Eq. (5-11), and the number of turns and their corresponding valve opening areas can be calculated in Table 5-1.

Table 5-1 NCCC needle valve opening and its corresponding turns

Point number	Q/L · min^{-1}	A_{vn}/m^2
Point 1, 1 turn	0.67	4.51×10^{-7}
Point 2, 2 turns	2.67	1.80×10^{-6}
Point 3, 3 turns	5.33	3.61×10^{-6}
Point 4, 4 turns	8.25	5.58×10^{-6}

5.2.2.4 Overall of the valve model

This section reveals the major calculation processes and equations that link all the valve models together, two sets of equations are utilised when the cylinder is extending or retracting.

When the asymmetric cylinder in the extending state:

$$\text{Piston side chamber}\begin{cases} Q_{D1} = A_1 v \\ Q_{C1} = Q_r - Q_{D1} \\ p_1 = \dfrac{Q_{C1} B}{V_1 s} \end{cases} \tag{5-12}$$

$$\text{Rod side chamber}\begin{cases} Q_{D2} = A_2 v \\ Q_2 = C_D A_{vn} \sqrt{\dfrac{2}{\rho}} \sqrt{p_2} \\ Q_{C2} = Q_{D2} - Q_2 \\ p_2 = \dfrac{Q_{C2} B}{V_2 s} \end{cases} \tag{5-13}$$

When the asymmetric cylinder in the retracting state:

$$\text{Piston side chamber}\begin{cases} Q_{\mathrm{D1}} = A_1 v \\ Q_1 = C_{\mathrm{D}} A_{\mathrm{vn}} \sqrt{\dfrac{2}{\rho}} \sqrt{p_1} \\ Q_{\mathrm{C1}} = Q_{\mathrm{D1}} - Q_1 \\ p_1 = \dfrac{Q_{\mathrm{C1}} B}{V_1 s} \end{cases} \tag{5-14}$$

$$\text{Rod side chamber}\begin{cases} Q_{\mathrm{D2}} = A_2 v \\ Q_{\mathrm{C2}} = Q_{\mathrm{r}} - Q_{\mathrm{D2}} \\ p_2 = \dfrac{Q_{\mathrm{C2}} B}{V_2 s} \end{cases} \tag{5-15}$$

The valve model in Matlab Simulink is depicted in Fig. 5-17.

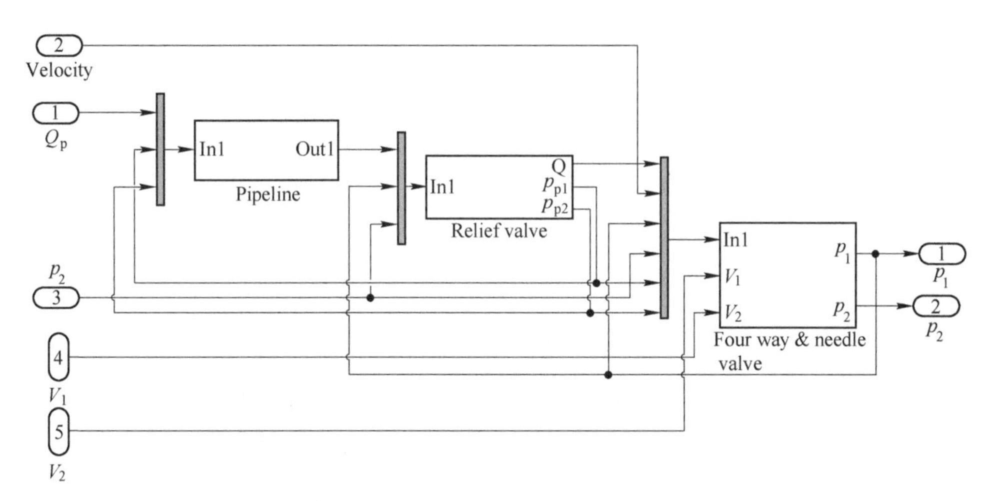

Fig. 5-17 Valve model in Simulink

It can be seen that the inputs to the valve model are the pump flow rate Q_{p} , velocity v , chamber pressures p_1 and p_2, chamber volumes V_1 and V_2, its outputs are the chamber pressures p_1 and p_2.

The valve model block diagram is depicted in Fig. 5-18, the four-way valve offers the system state in the extending or retracting state.

5.2.3 Actuator model

This actuator model is to calculate the net force applied on the load, it accepts inputs of friction force F_{f} , chamber pressures p_1 and p_2. Substitute them into below force balance

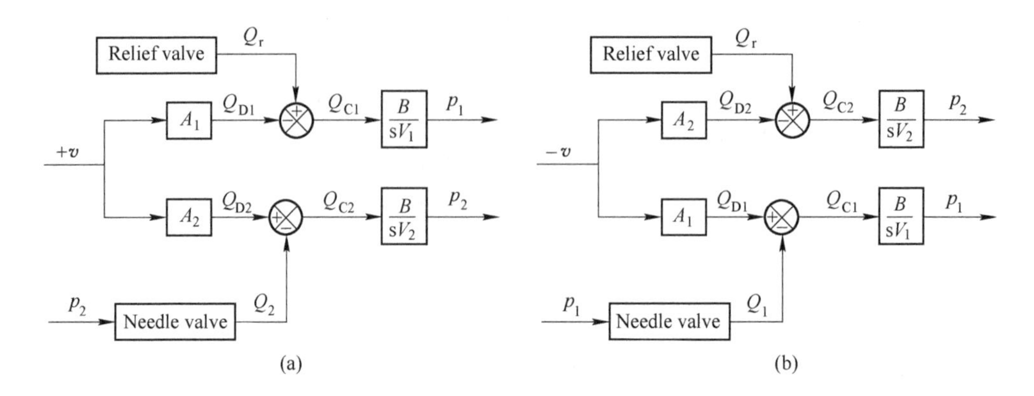

Fig. 5-18 Valve model in extending state (a) & retracting state (b)

equation:

$$F_{net} = p_1A_1 - p_2A_2 - F_f \tag{5-16}$$

The actuator model outputs the net force F_{net} to the load, and its model in Matlab Simulink are depicted in Fig. 5-19.

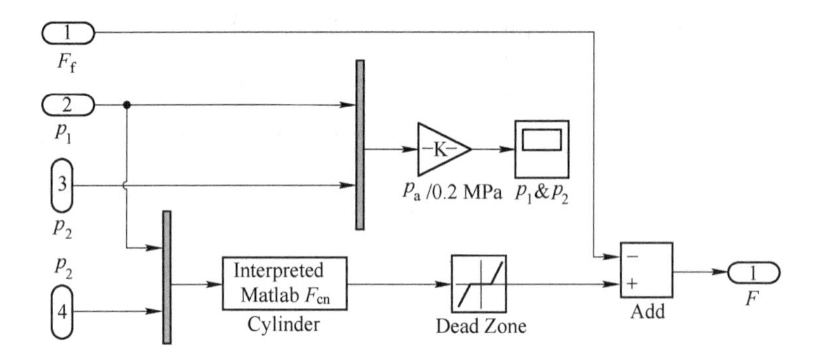

Fig. 5-19 Actuator model in Matlab Simulink

The actuator model block diagram is depicted in Fig. 5-20.

5.2.4 Load model

The load model receives the net force F_{net} from actuator model and chamber pressures p_1 and p_2 from valves model, the chamber pressures are used to calculate the friction force with the new friction model in Chapter 4, the net force applied on the load generates the acceleration

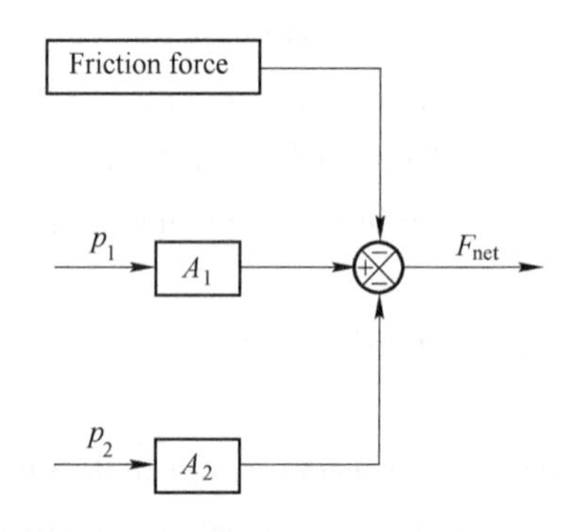

Fig. 5-20 Actuator model block diagram

a by:

$$a = \frac{F_{net}}{m} \tag{5-17}$$

The velocity v is calculated by the integration of the acceleration a, assuming initial condition of $a=0$.

$$v = a \cdot \frac{1}{s} \tag{5-18}$$

The cylinder displacement l is calculated by the integration of velocity v, assuming initial condition of $v = 0$.

$$l = v \cdot \frac{1}{s} \tag{5-19}$$

With the displacement l, the cylinder chamber volumes V_1 and V_2 can be calculated by:

$$V_1 = (0.15 + l)A_1 \tag{5-20}$$

$$V_2 = (0.15 - l)A_2 \tag{5-21}$$

So that the load model outputs the velocity v, friction force F_f, chamber volume V_1 and V_2, and this model is depicted in Fig. 5-21.

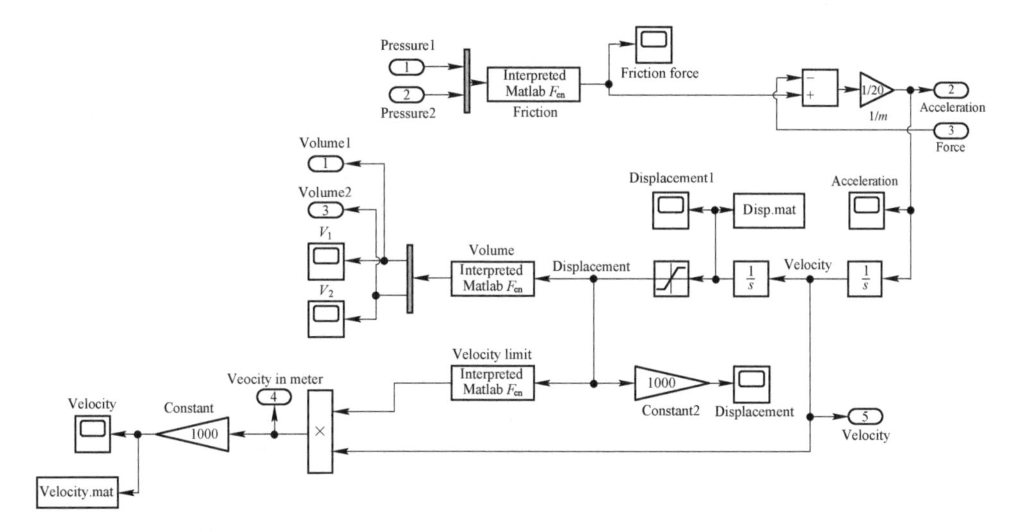

Fig. 5-21 The load model in Matlab Simulink

The load model block diagram is depicted as in Fig. 5-22.

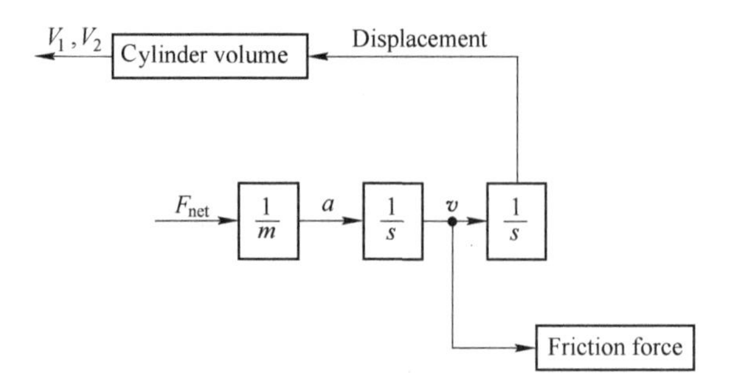

Fig. 5-22 The load model block diagram

5.3 The final model

Combine all the models in section 5.2 to generate the overall open loop model of the system as in Fig. 5-23, which reflects that this model still utilises the algorithm of the component linking method. But due to there is no constant supply pressure to drive the cylinder, the asymmetric cylinder chamber pressures p_1 and p_2 must be calculated separately in cylinder extending and retracting state, the key equations are still referred to the work done by Leaney[63]. The simulation results will be obtained and discussed in the next chapter.

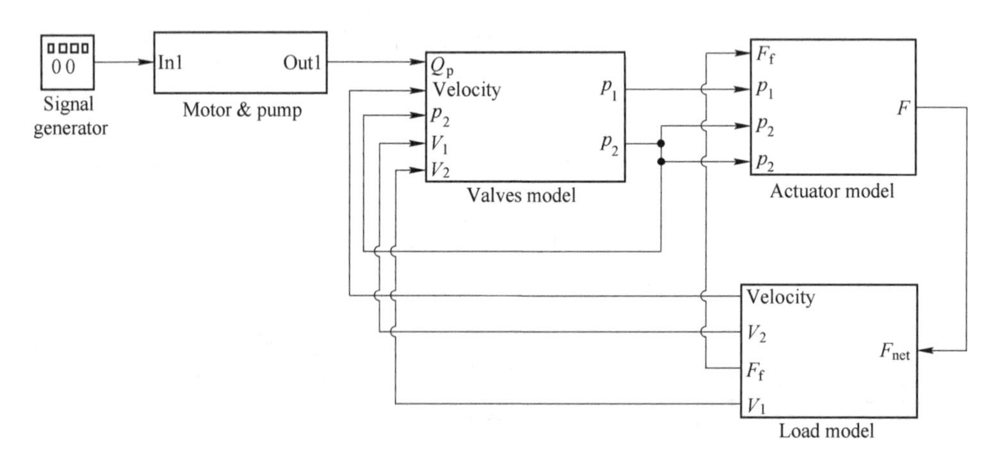

Fig. 5-23 The hybrid pump controlled asymmetric cylinder drive system in Matlab Simulink

The overall block diagram when the system in the extending state is depicted in Fig. 5-24.

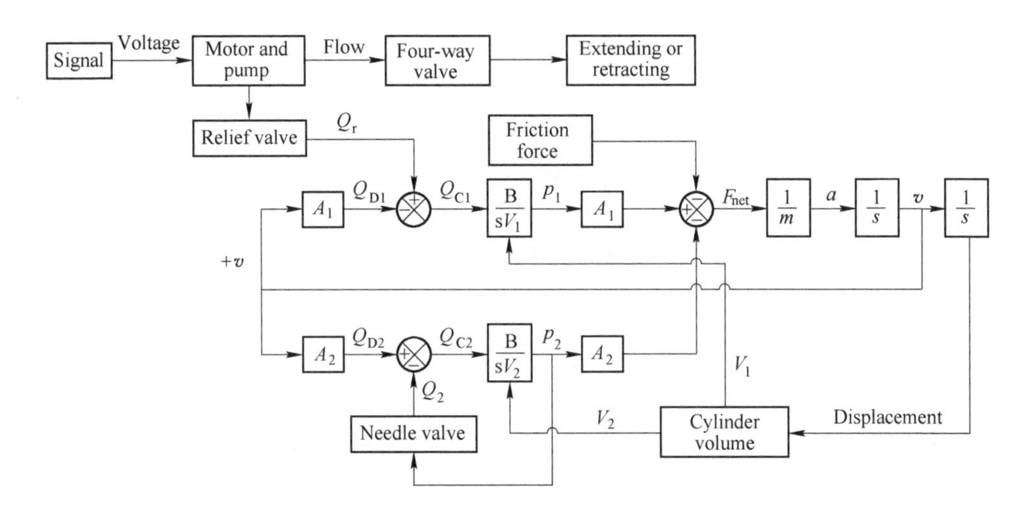

Fig. 5-24 Block diagram of hybrid pump-controlled asymmetric cylinder system when in the extending

When the system in the retracting state, simply replace the extending valve model with retracting valve model in Fig. 5-18 (b) to model the system in retracting state.

Chapter 6 Experiment to Explore the Truth, Simulation to See the Future

To validate the model proposed in Chapter 6 and move forward the nonlinear research of the hybrid pump-controlled asymmetric cylinder drive system, open loop simulation and experimental results of the test rig are utilised for nonlinearities analysis and model validation.

The experimental tests are carried out in an open loop to explore the system nonlinear behaviours and performance, the simulation results are validated by the experimental results, so that the model can be used to understand the causes of these nonlinearities. As a result, the validated system model can be utilised in energy comparison argument in the next chapter.

6.1 Square wave test results analysis

To identify the hybrid pump-controlled performance under a steady state, the square wave tests are carried out. To ensure the system operates in a steady state, a low-frequency square wave of 0.2 Hz command is sent to the system.

The command was sent to the motor drive as the voltage form, then the pump operates at a certain velocity with a certain ratio. As the structure of the asymmetric cylinder, the piston side area A_1 is larger than the rod side area A_2. When the same flow rate is delivered to the cylinder chambers, the cylinder will be retracting faster than extending. So that a bias is required to add to the command to perform a symmetric motion on the asymmetric cylinder.

A biased command is sent to the motor drive to achieve a ±50 mm/s square wave motion on the cylinder without added load, its velocity responses are described as in Fig. 6-1.

The experimental results can be observed with a lot of spikes, this may be caused by multiple reasons. The velocity is obtained by the derivative of the displacement data

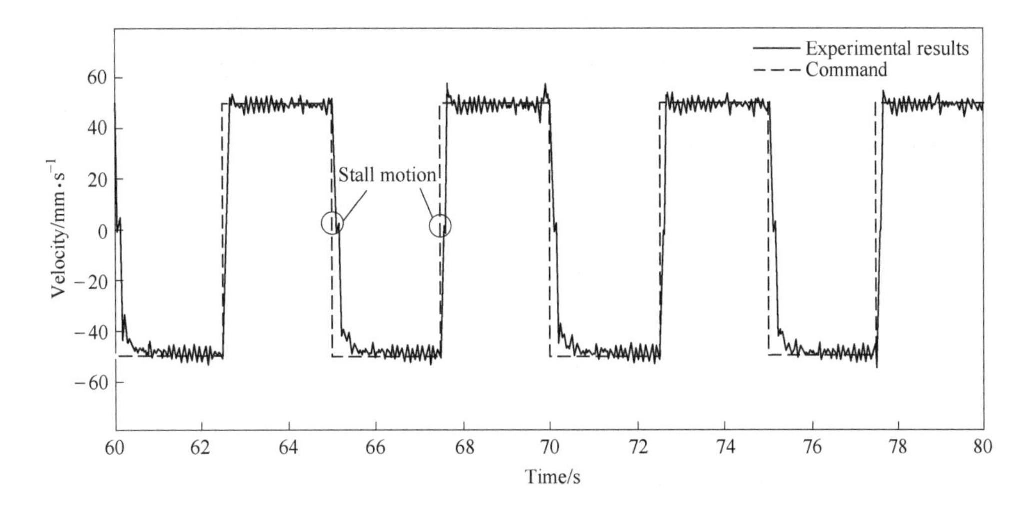

Fig. 6-1 Experimental results of ±50 mm/s square wave cylinder motion without added load

recorded by a low-cost potentiometer, the displacement data is processed with a low pass filter to attenuate the noise.

The low pass filter suppresses most of the high-frequency background noise with a cut off frequency of 5 Hz, but still leaves some oscillations in displacement data, hence after the integral such oscillation is amplified as velocity spikes in Fig. 6-1. The potentiometer is attached to the cylinder by tapes, the vibration may also cause the spikes during recording.

It can be noticed that when the asymmetric cylinder in the extending state, the velocity responds faster to the command than that in the retracting state. As the cylinder is driven by the chamber pressures p_1 and p_2, such a phenomenon must relate to the pressures. Observe the chamber pressures measured by the pressure transducer as in Fig. 6-2.

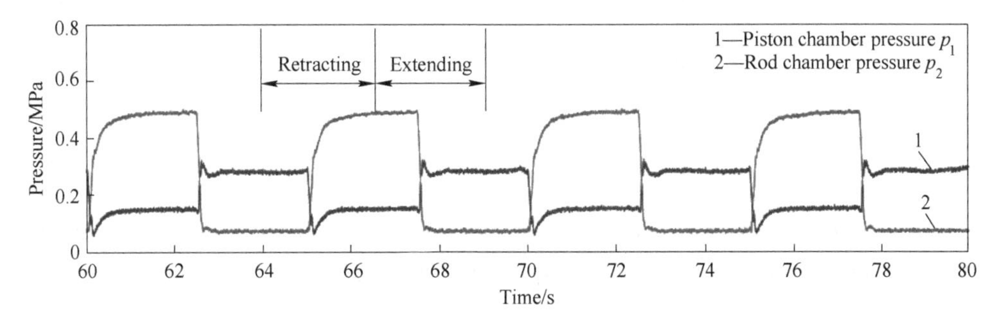

Fig. 6-2 Asymmetric cylinder chamber pressures p_1 and p_2 when in ±50 mm/s square wave motion without added load

The same phenomenon is observed as in the above figure, the pressure responses when the cylinder in the extending is faster than that in the retracting. This difference can be interpreted by the design of the return line in the hybrid pump controlled asymmetric cylinder drive system as in Fig. 6-3.

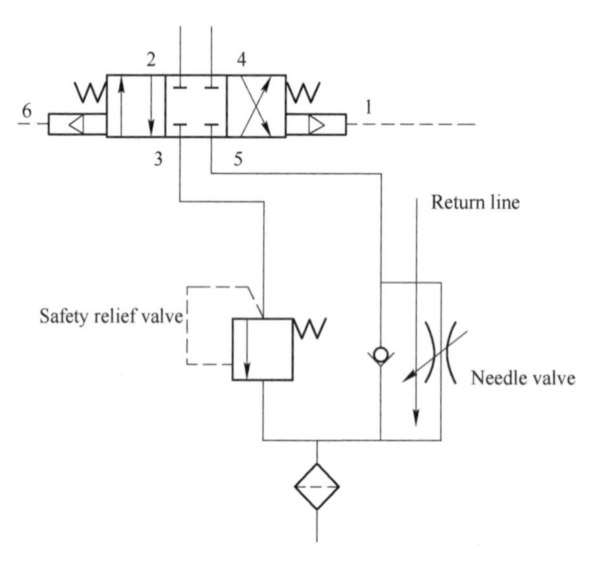

Fig. 6-3 Return line of the hybrid pump controlled asymmetric cylinder drive system

The safety relief valve is set at 20 MPa pre-setting, so that it is closed during normal system operation. The needle valve is manually adjusted and remains the same no matter cylinder extending or retracting, so that flow pressure equation for the return line can be written as:

When extending in steady state:

$$Q_{\text{return}} = A_2 v = C_D A_{\text{vn}} \sqrt{\frac{2}{\rho}} \sqrt{p_2} \tag{6-1}$$

where A_{vn} is the valve opening area of the manually adjusted needle valve, m^2.

When retracting in a steady state:

$$Q_{\text{return}} = A_1 v = C_D A_{\text{vn}} \sqrt{\frac{2}{\rho}} \sqrt{p_1} \tag{6-2}$$

With a certain cylinder velocity v, there existing corresponding chamber pressures p_1 and p_2. Rearrange the Eq. (6-1) and Eq. (6-2) reveals:

$$\frac{A_2}{A_1} = \frac{\sqrt{p_2}}{\sqrt{p_1}} \tag{6-3}$$

The p_2 in Eq. (6-1) is the rod side chamber pressure steady state value when cylinder extending, and the p_1 in Eq. (6-2) is the piston side chamber pressure when cylinder in the retracting state. So that when the cylinder chamber is connected to the return line, the piston chamber pressure p_1 will always be larger than the rod side chamber p_2.

Meanwhile, in order to achieve the force balance in a steady state (neglect the friction force here for the ease of analysis), so that:

When extending $p_{1\_extending}A_1 = p_{2\_extending}A_2$

When retracting $p_{1\_retracting}A_1 = p_{2\_retracting}A_2$

And based on $\frac{A_2}{A_1} = \frac{\sqrt{p_{2\_extending}}}{\sqrt{p_{1\_retracting}}}$

Revealing that when the cylinder extends and retracts at the same velocity, the $p_{2\_retracting}$ is much larger than the $p_{1\_extending}$, this is a major nonlinear character of the asymmetric cylinder drive system. Hence, due to this character, when retracting the pump line needs more time to build a higher pressure like $p_{2\_retracting}$, so that the time constant of pressure response is larger, leading to the phenomenon of slower velocity response when retracting like in Fig. 6-1.

Another phenomenon in the square wave test is stall motion as labelled in Fig. 6-1, the cylinder will be stationary for a short amount of time when the cylinder switches its state from extending to retracting or from retracting to extending. Zoom in the stall motion as in Fig. 6-4.

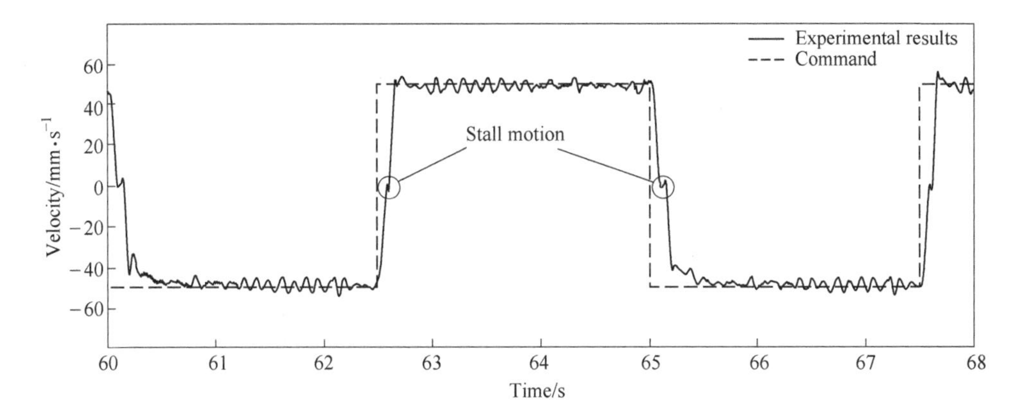

Fig. 6-4 Zoom in the stall motion when the cylinder switch its state

The stall motion occurs in every period of square wave motion, though the stall motion in Fig. 6-4 should be stationary (a short horizontal line), it appears as a jerky oscillation, which may be caused by the potentiometer that is attached to the cylinder by tapes. Such setup is not perfectly stable and can lead to such a problem. During experimental operation, when the cylinder changes its direction, it will be stationary for a very short amount of time and then move to its target direction. In theory, there should be two factors that cause this phenomenon:

(1) Stiction friction;

(2) Pressure charging process in the pipeline.

The stiction friction happens when the velocity is low, and it is relatively larger than the pressure difference on both sides of the cylinder piston. However, when the cylinder is stalled, the pressure in the pipeline keeps accumulating and increasing until the pressure difference p_e overcomes the stiction friction.

Another factor is the pressure charging process, when the pump changes its operation direction, the pressure in pipeline needs to be recharged to drive the pilot-shifted four-way valve to create a pathway. For instance, when the pump changes its direction from counterclockwise to clockwise as in Fig. 6-5, the charge pressure in the retracting line starts to decline. Meanwhile, the charge pressure in the extending line starts to accumulate. This process is short, but due to the design of this circuit, the stall motion in Fig. 6-4 is inevitable.

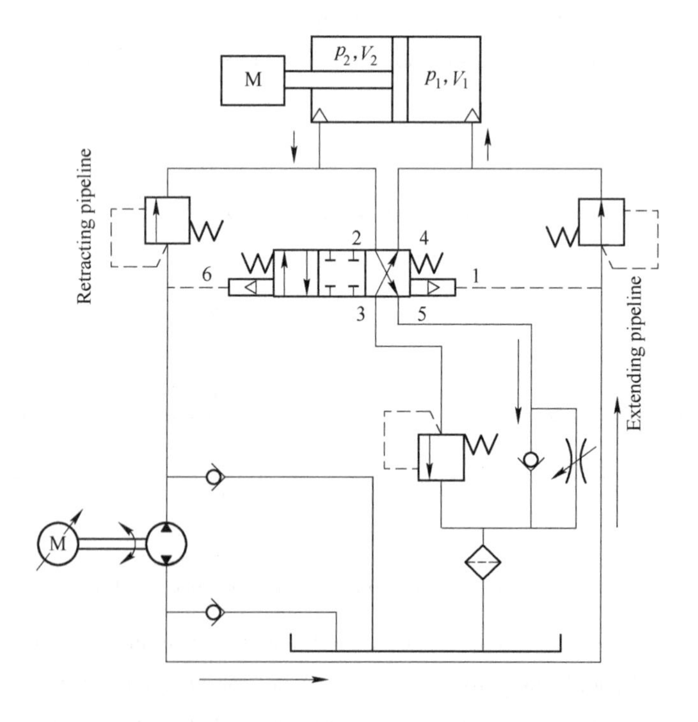

Fig. 6-5 Cylinder change its direction from retracting to extending

Based on the difference of response time constants in the extending and retracting, the cylinder displacement in this ±50 mm/s symmetric velocity square wave test also differs from prediction. The asymmetric cylinder displacement test results are depicted as in Fig. 6-6.

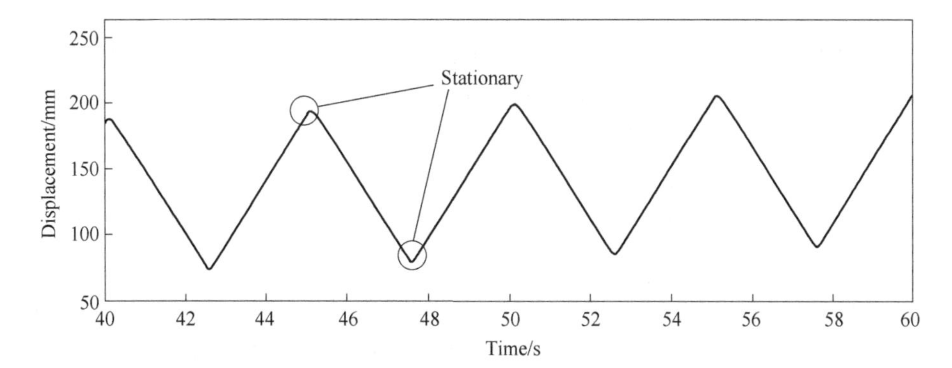

Fig. 6-6 Asymmetric cylinder displacement test results under ±50 mm/s square wave motion

Two phenomena can be found from Fig. 6-6, the first one is the same as the stall motion in Fig. 6-4, a short time of cylinder stationary occurs at each of the displacement curve peak and bottom place. The other one is the trending of extension, the cylinder moves back and forth, but it is gradually extending to the cylinder stroke end.

The cylinder stationary can be interpreted the same as the reason of stall motion in Fig. 6-4. The trending of extending can be explained within the Fig. 6-7, the shadow area in it is the difference between the command and experimental velocity results. There is less shadow area in the extending than that in the retracting state.

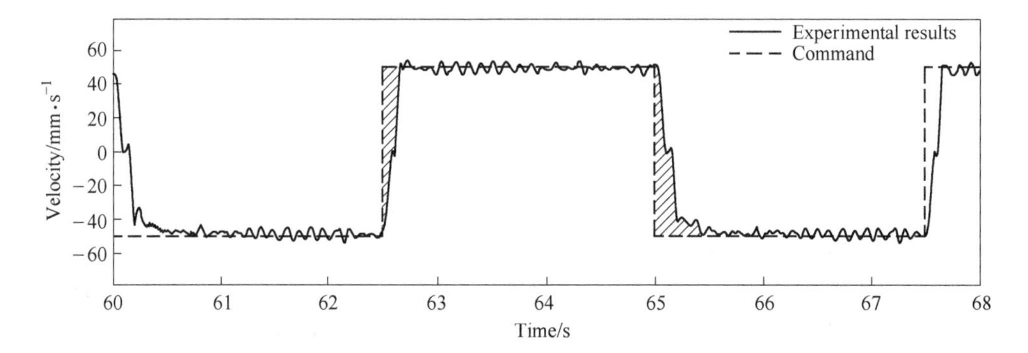

Fig. 6-7 The error area between command reference and experimental results

If the system is able to follow the command reference velocity perfectly, the cylinder displacement will back and forth perfectly without bias to extending or retracting end. However, a perfect system does not exist, errors will occur between the outcome and command reference. When the cylinder is retracting, the velocity response time constant is larger, so that the error area in the retracting state is larger than that in the extending state. As a result, under the symmetric square wave velocity command, the asymmetric cylinder will gradually extend to the stroke end.

In Chapter 4, a new friction model is proposed, in which the friction force is not only related to the velocity, but also the chamber pressure difference and acceleration. The experimental friction results are obtained by:

$$p_1A_1 - p_2A_2 - m \cdot a = \text{friction force} \tag{6-4}$$

where a is the acceleration, which is obtained by the derivative of the measured velocity.

The experimentally measured friction force results are depicted in Fig. 6-8, similar to the velocity results and chamber pressure results, the friction force results also have a smaller time constant when getting into extending state than that getting into retracting state. This difference is due to the friction in this hybrid pump-controlled asymmetric cylinder drive system is related to the chamber pressure difference, and the time constant of chamber pressures response is smaller in the retracting state as in Fig. 6-2.

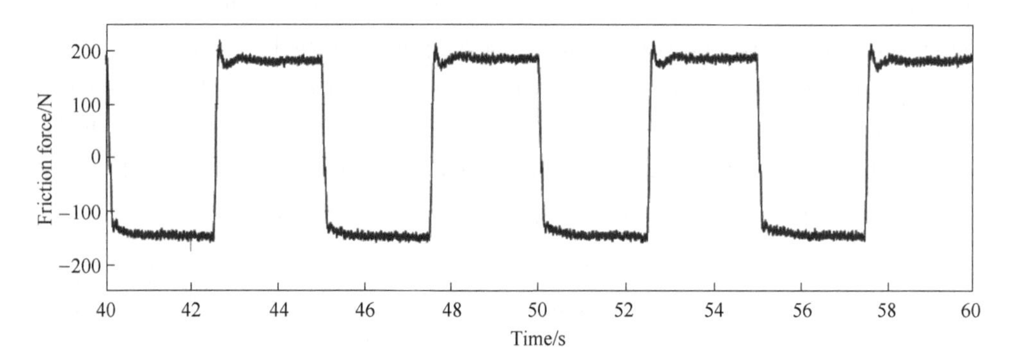

Fig. 6-8 Measured friction force under cylinder ±50 mm/s square wave motion without added load

Meanwhile, though the test velocity is symmetric ±50 mm/s, the absolute value of friction force in the retracting is noticeably smaller than that in the extending. This phenomenon can be interpreted by Eq. (6-4), when the cylinder in the extending or retracting steady state, the acceleration a in Eq. (6-4) is zero. Though the absolute pressure difference $abs(p_1 - p_2)$ in the retracting is larger, with the influence of piston and rod side area A_1 and A_2, the absolute value of $p_1A_1 - p_2A_2$ is still smaller in the retracting.

These square wave experimental results describe the nonlinearities of the hybrid pump-controlled asymmetric cylinder drive system, there are some similar characters to the valve-controlled asymmetric cylinder drive system, like the pressure jump when cylinder switch its state, this nonlinearity is due to the structure of asymmetric cylinder. Some other nonlinear behaviours like stall motion and different time constant are unique in the hybrid pump-controlled system. The square wave tests are to explore the system performance in a steady state, and its dynamic performance will be discussed in the next section.

6.2 Sine wave test results analysis

To identify the dynamic performance of the hybrid pump-controlled asymmetric cylinder drive system, the sine wave test is implemented. A sine wave command signal is sent to the pump to drive the velocity of the asymmetric cylinder in a sine wave motion.

Same as the square wave test, due to the asymmetric cylinder structure, the command signal is biased to achieve the cylinder symmetric motion. The cylinder velocity test results are described as in Fig. 6-9, it is commanded to achieve a 0.2 Hz±50 mm/s sine wave motion.

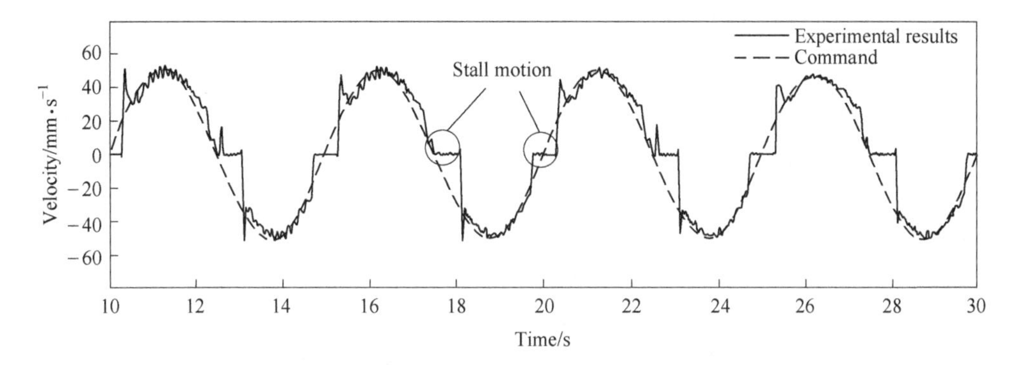

Fig. 6-9 ±50 mm/s sine wave velocity motion test results

The most noticeable phenomenon is the stall motion time labelled in Fig. 6-9 is much larger than that in square wave motion. There are two factors that affect this character:

(1) Stiction friction;

(2) Pressure charging process in the pipeline.

The reason is the same as that in square wave motion. When the cylinder velocity switches its direction, the square wave motion increases velocity to a certain value almost instantaneously, but the sine wave gradually increases its velocity to peak value. So that the motor charges the pipeline slower when in sine wave motion, leading to a larger amount of time of zero velocity gap in Fig. 6-9.

The system components behaviours of the stall motion can be analysed by the performance of chamber pressures during operation as in Fig. 6-10, it is more complicated than the chamber pressures when cylinder in a square wave motion.

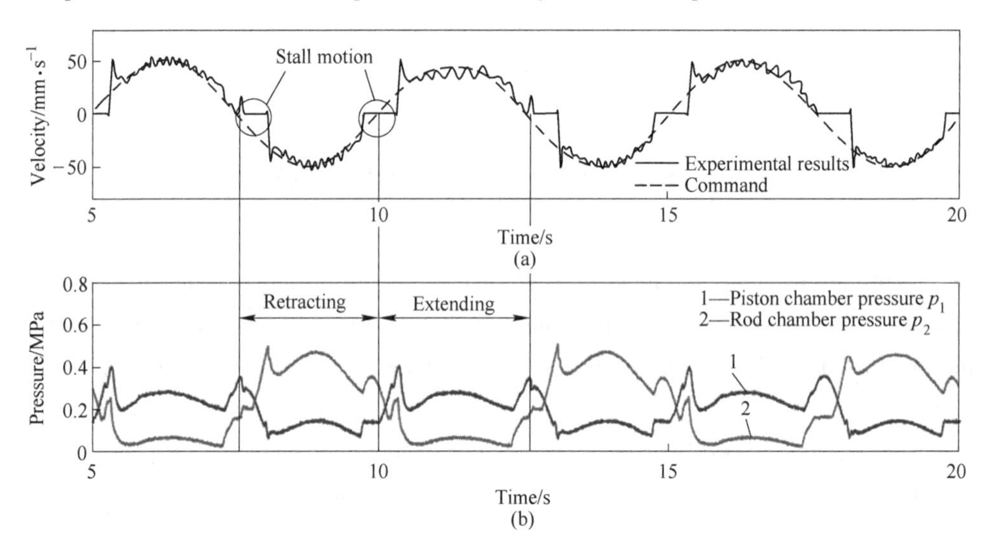

Fig. 6-10 The chamber pressures p_1 and p_2 in cylinder sine wave ±50 mm/s motion without added load

(a) Velocity & command; (b) Chamber pressures

Overall, the chamber pressures performances' characteristics are similar between the sine wave and square motion, the piston chamber pressure p_1 when cylinder in the retracting state is larger than the rod chamber pressure p_2 when cylinder in the extending state. Due to the force balance, the rod chamber p_2 when cylinder in reacting is much larger than the piston chamber pressure p_1 when cylinder in the extending state.

The velocity stall motion in sine wave test in Fig. 6-10 can be analysed as:

(1) The first labelled stall motion zone occurs after cylinder getting into retracting state, a short time before this stall motion, chamber pressures p_1 and p_2 increasing until the stall motion starts. This can be interpreted as that the four-way valve in the system returning to its neutral position, relief valve 1 opened and relief valve 2 closed. The pump still delivers fluid into the piston chamber and the fluid in the rod side chamber cannot be released. So that both chamber pressures p_1 and p_2 are increasing in this state. When the pump starts to deliver fluid into rod side chamber (in reacting state), the four-way valve in the retracting state, the relief valve 1 closed and relief valve 2 opened as in Fig. 6-11 (b). So that the chamber pressure p_2 starts to increase and the

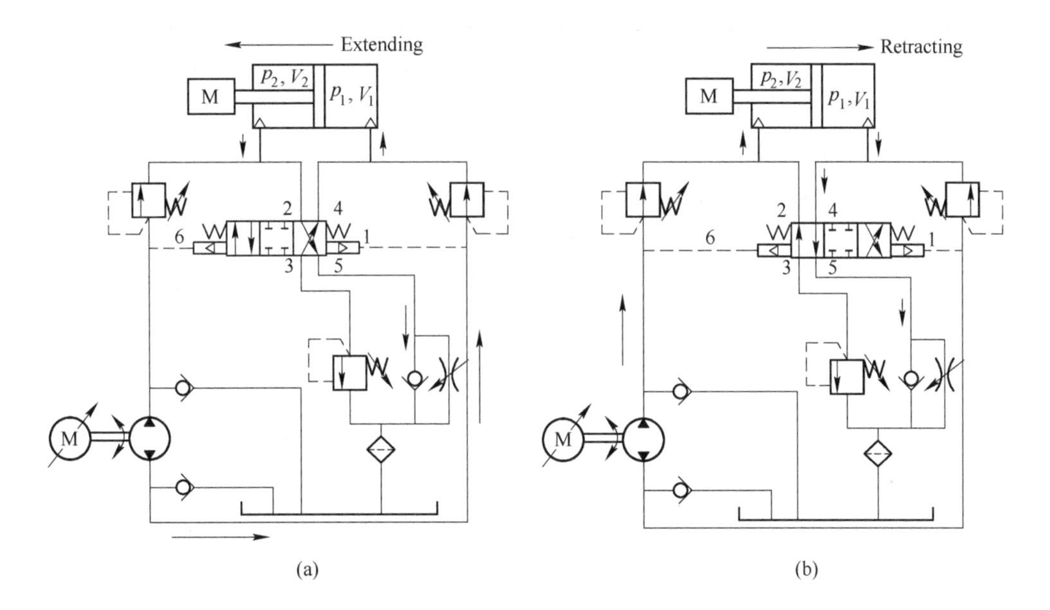

Fig. 6-11 Proposed hybrid pump controlled asymmetric cylinder drive system in extending (a) and retracting (b) state

fluid in the piston chamber can be released to the tank.

Due to the existence of friction, the cylinder stays stationary until the chamber pressure difference is large enough to overcome the friction force.

(2) The second labelled stall motion occurs when the cylinder at the end of the retracting state and the start of extending state. At the end of retracting state, the four-way is in the neutral position, relief valve 1 is closed and relief valve 2 is opened. As the cylinder is in stationary, the stiction friction force is involved and chamber pressure p_2 is not large enough to overcome the friction force, so that the chamber pressure p_1 is not changed during the end of retracting. By then the four-way valve is still in the neutral position, relief valve 1 is opened and relief valve 2 is closed. The chamber pressure p_1 starts to increase due to the pump starts to deliver fluid into the piston side chamber, but the chamber pressure p_2 increases due to the rod side chamber is not connected to the tank.

At last, the four-way valve is in the extending state, the relief valve 1 is opened and relief valve 2 is closed as in Fig. 6-11 (a), the cylinder stays stationary until the chamber pressure difference overcomes the friction force.

The cylinder displacement experimental results when it is in ±50 mm/s is depicted in Fig. 6-12, it has the same trending of extending to the stroke end after it moves back and forth, the causes are the same as in the last section and not necessary to repeat here. Its stationary time interval is noticeably larger than that of square wave motion, it can be interpreted with the same factors:

(1) Stiction friction;

(2) The pressure charging process in the pipeline.

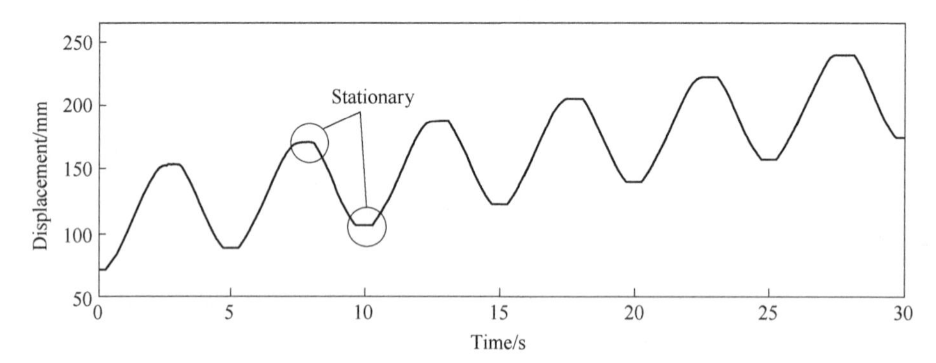

Fig. 6-12 Measured cylinder displacement when it in ±50 mm/s sine wave motion

As the character of the sine wave motion, when it is changing its direction, the velocity is slow and leads to the charging process requiring more time than that of square wave motion test.

The friction plays an important role in above nonlinearities, the value of friction force during this sine wave operation is depicted as in Fig. 6-13, the friction spikes occur at the stall motion zone, indicating the stiction friction force is one of the reasons that the cylinder will be stationary for a short time when it changes its state.

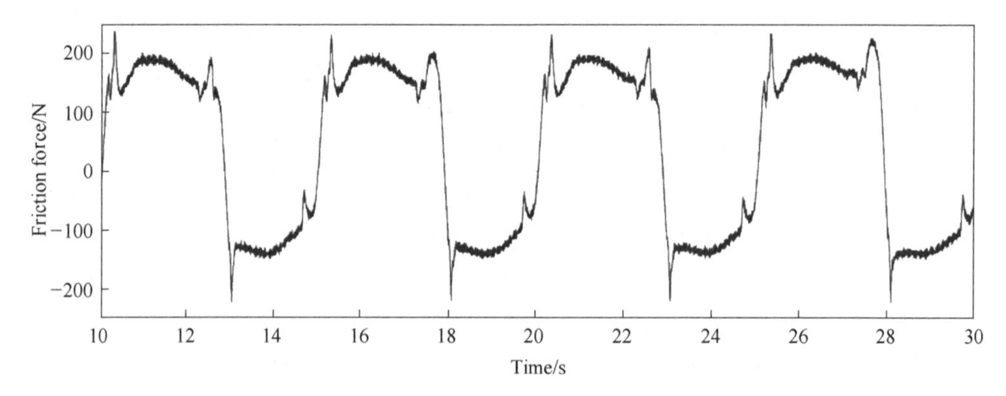

Fig. 6-13 Measured friction force when cylinder in ±50 mm/s sine wave motion

This section describes the experimental results when the hybrid pump-controlled asymmetric cylinder drive system performs the ±50 mm/s sine wave, the stall motion time in sine wave test is longer than that in square wave time. Based on the analysis of chamber pressures, the valve operation during cylinder stall motion is discussed. The measured friction force proves the stiction friction force is one of the reasons for the cylinder stall motion. The dynamic performance of the hybrid pump-controlled system is analysed in this section, these results will be used to validate the simulation model results in the next section.

6. 3 Simulation results

This section depicts the open loop simulation results of the system model in Chapter 5, they are compared with experimental results to validate its consistency, so that the model can be utilised in Chapter 8 and further research.

6. 3. 1 Square wave simulation results

Starts with square wave simulation, the ± 50 mm/s velocity simulation results are compared with experimental results as in Fig. 6-14.

The stall motion in simulation is easier to be identified, the different velocity response time constant is captured by the simulation model.

As the chamber pressures drive the cylinder to accomplish its target motion, the simulation results of chamber pressures are compared with experimental results as in Fig. 6-15.

Overall, the simulation results capture the behaviours of experimental chamber pressures, the difference in simulation is the result of numerical calculation which will occur when the valve switches its direction.

As proposed by the new friction model in Chapter 4, the friction on the cylinder piston is not only affected by velocity, but also the chamber pressure difference. The friction simulation results of the new friction model based on the simulation chamber pressures and velocity are compared with experimentally measured friction force as in Fig. 6-16. The new friction model simulation result shows consistency with the experimental result.

This section validated the simulation model performance when the hybrid pump-controlled asymmetric cylinder drive system in square wave motion, some numerical calculation leads to the spikes in Fig. 6-16, but the overall model simulation results

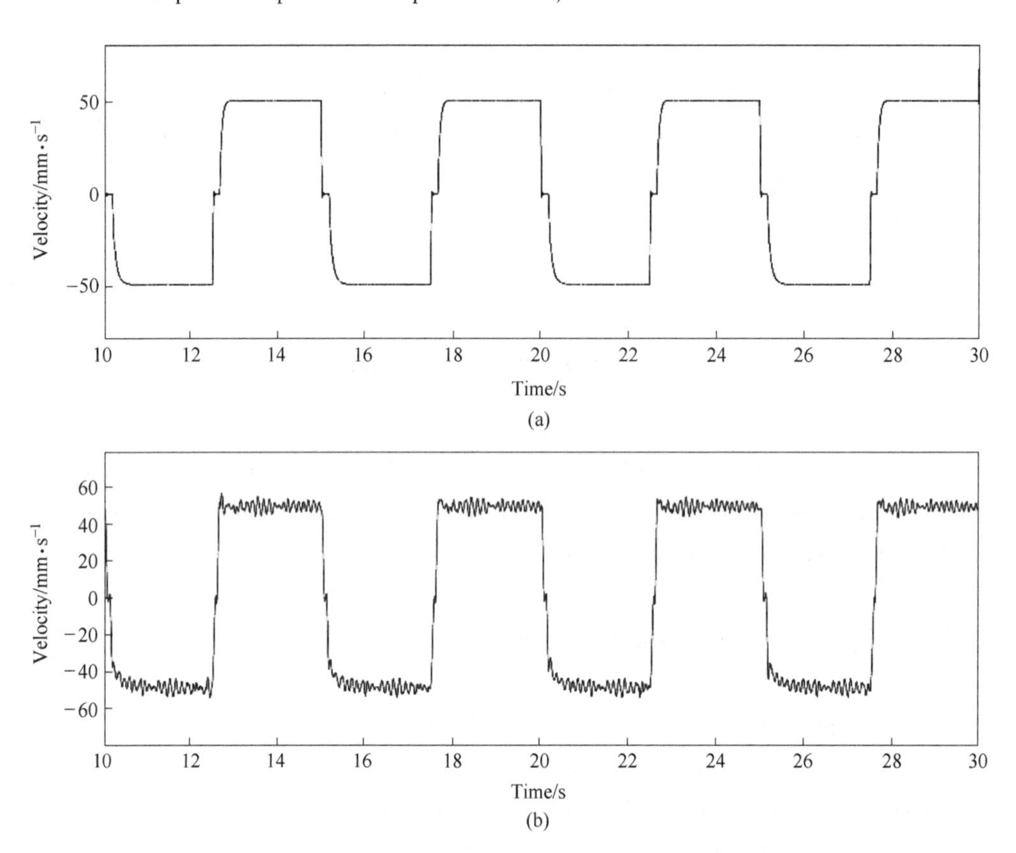

Fig. 6-14 Square wave velocity simulation (a) and experimental (b) results (±50 mm/s square wave)

show the consistency with the experimental results.

So that the simulation model proposed in Chapter 5 is validated when the system is operating in a steady state, the dynamics performance of the model will be validated in the next section.

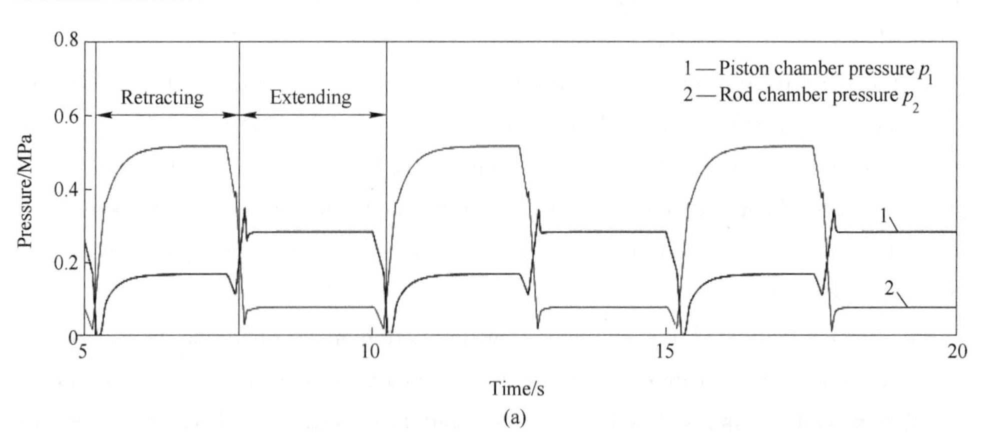

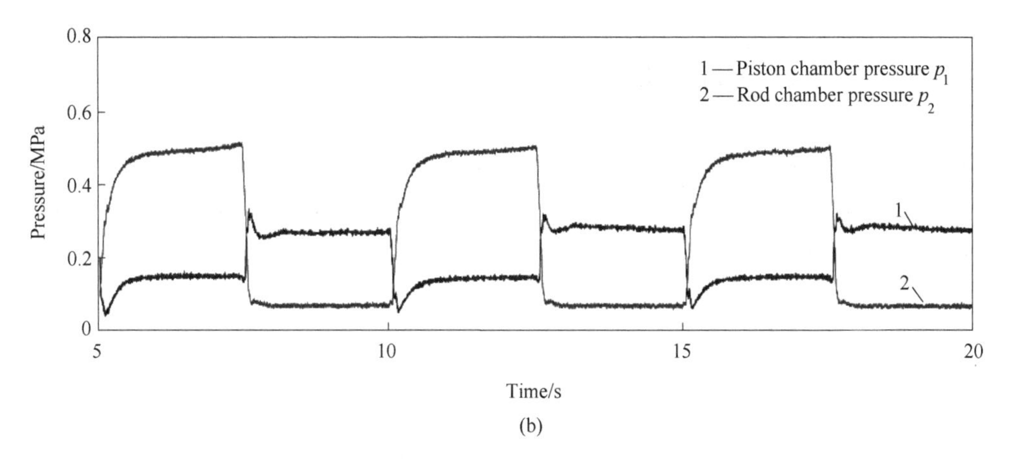

(b)

Fig. 6-15 Square wave chamber pressures simulation results (a) and experimental results (b) (±50 mm/s square wave)

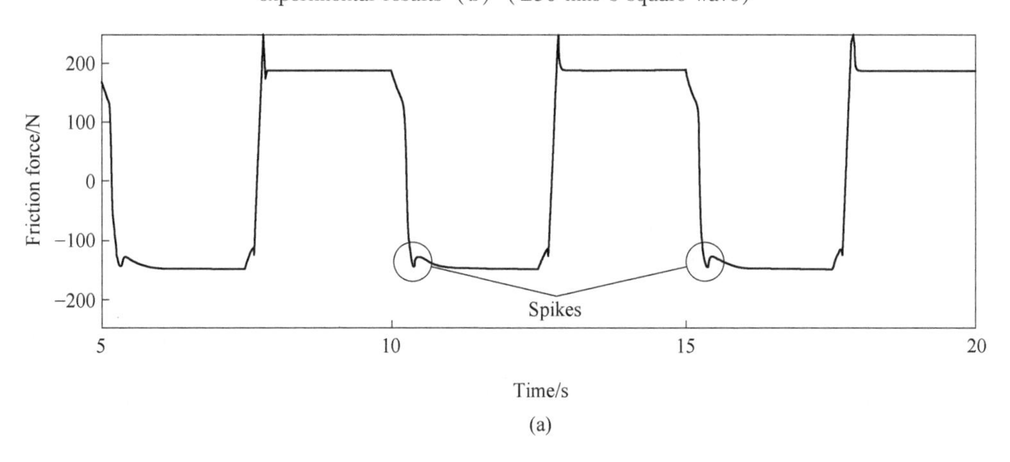

(a)

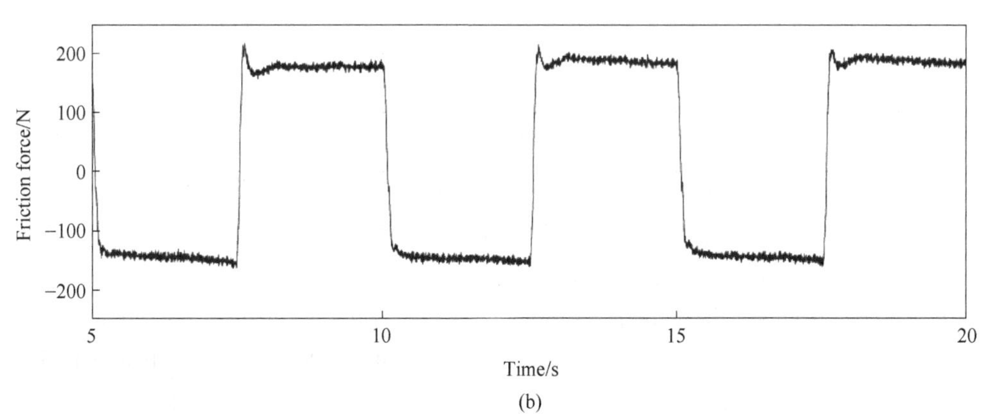

(b)

Fig. 6-16 Friction force simulation results (a) and experimental results (b) when cylinder in square wave motion (±50 mm/s square wave)

6.3.2 Sine wave simulation results

This section describes and analyses simulation results when the asymmetric cylinder is operating a ±50 mm/s sine wave motion, the simulation results and experimental results are compared as in Fig. 6-17.

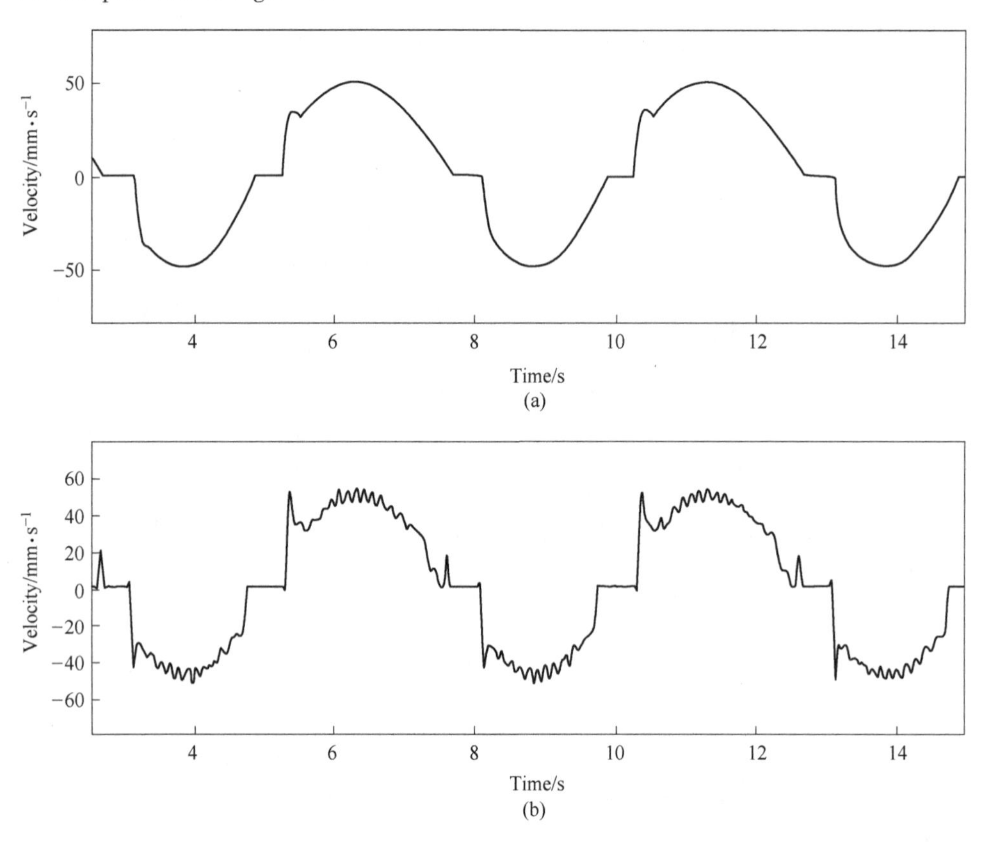

Fig. 6-17 Velocity simulation (a) and experimental (b) results when cylinder in sine wave motion

Similar to the square wave test, the velocity oscillation is caused by the measurement data from the low-cost potentiometer. The velocity spike when cylinder switches from retracting to extending is captured by the simulation model, but not as obviously as experimental results, the larger stall motion when cylinder changes its direction in experiment test is simulated by the model. Though the velocity spike when cylinder switches from extending to retracting in the experimental test is not captured by simulation results, it can be regarded as a transient effect and not affect the overall model performance consistency with the experimental results.

The chamber pressures are the power source to drive the asymmetric cylinder, the velocity response of the hybrid pump-controlled system is complicated as in Fig. 6-17, the experimental chamber pressure results should describe more unexpected dynamics. The system chamber pressures simulation and experimental results are depicted as in Fig. 6-18.

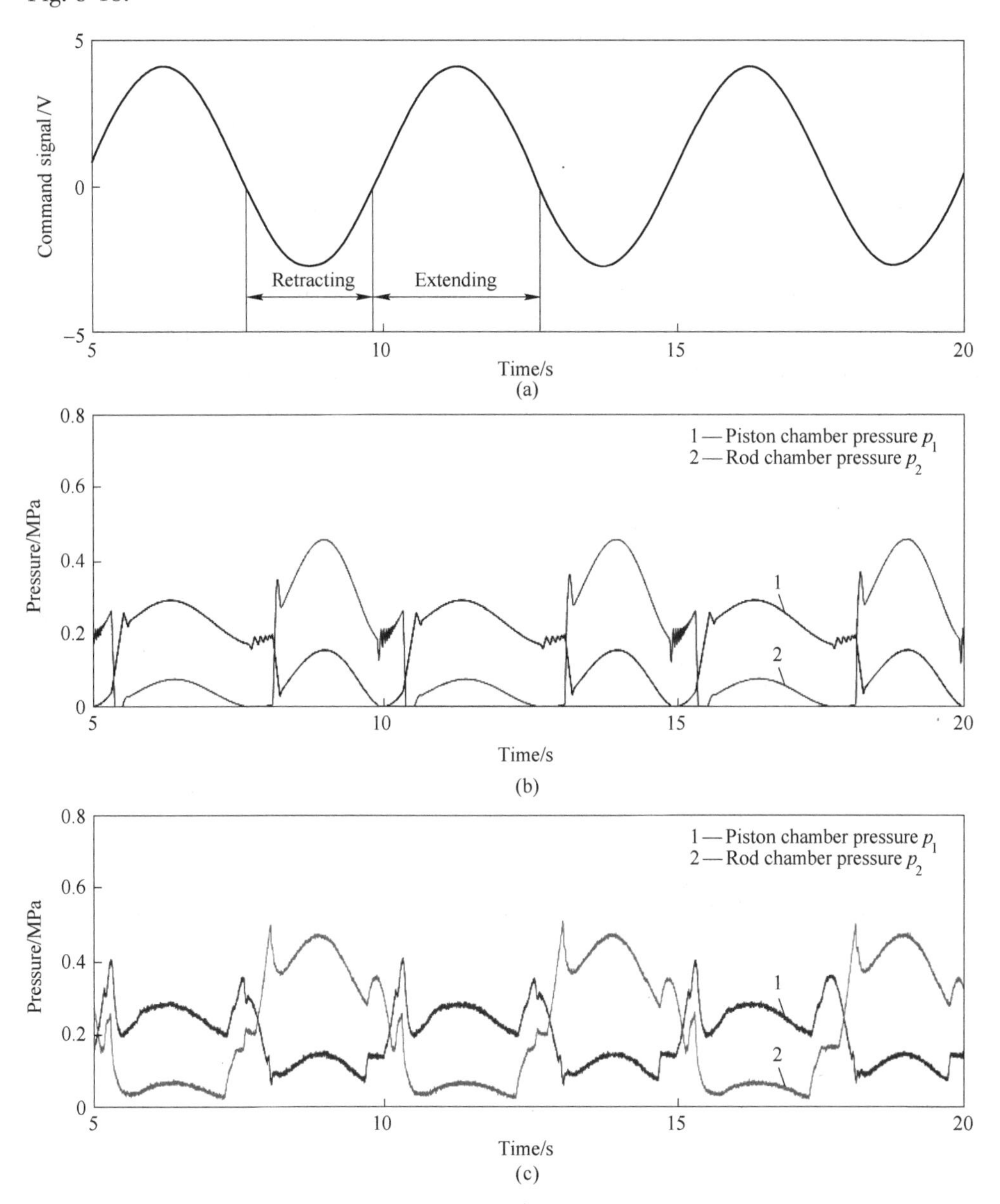

Fig. 6-18 Simulation and experimental results of chamber pressures when cylinder in sine wave motion
(a) Command signal; (b) Simulated pressures; (c) Experimental measured pressures

The chamber pressures simulation results capture most of the dynamics of the experimental results, the major difference occurs when the stall motion appears, and this is related to the valves switching mechanism in the hybrid pump-controlled asymmetric cylinder drive system. The simulation model is to simulate the realistic system as closely as possible, there is no perfect model that able to predict all the performance of the system. Overall, the consistency of simulation results and experimental is acceptable.

Friction force is one of the most important dynamic responses affecting low-velocity performance in this system, the simulation results and experimental measurements are compared as in Fig. 6-19. The dynamic behaviours before and after the velocity reversal are simulated by the model, but the value of friction force spikes as labelled in Fig. 6-19 is not perfectly captured by the simulation model. The overall simulation performance shows consistency with experimental results, the spikes can be caused by the valves switching operations, friction force influence at low velocity or combined.

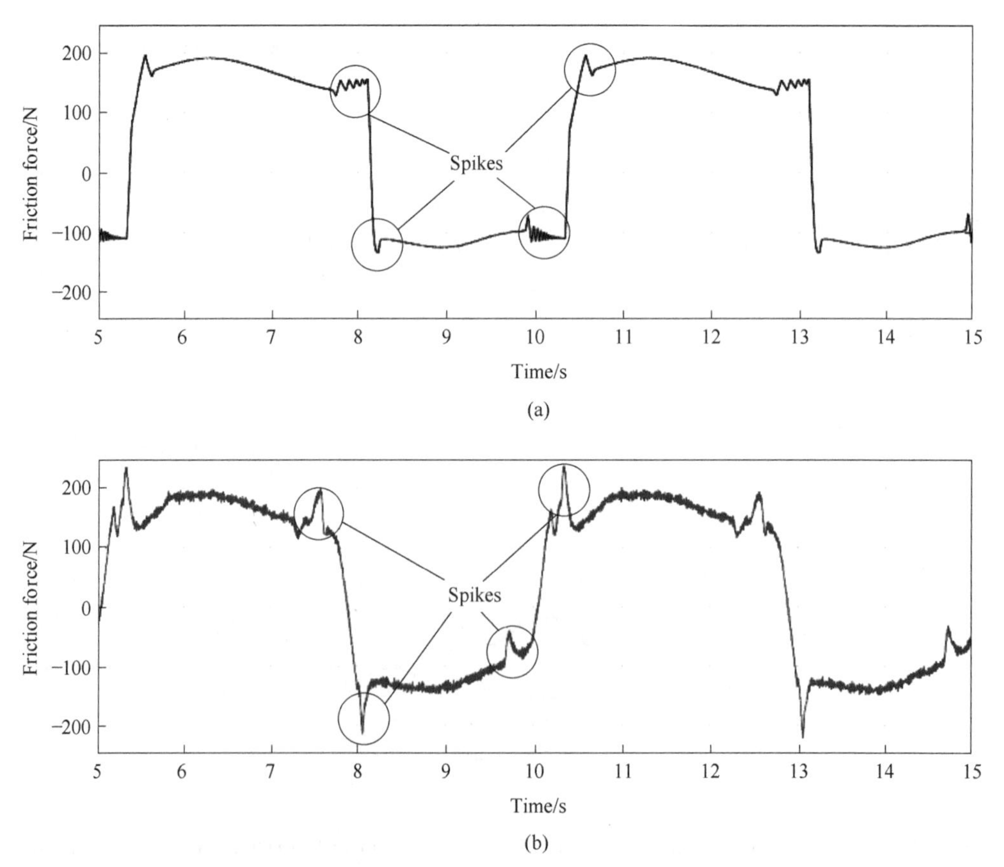

Fig. 6-19 Friction force simulation (a) and experimental (b) results when cylinder in sine wave motion

As the new friction model in this simulation model utilised the simulated velocity, chamber pressures and accelerations from simulation results, due to these differences between simulation and experimental results, the modelled friction force results depicted in Fig. 6-19 do not indeed indicate the accuracy of the new friction model. If the friction model directly utilises the data from experimental results like in Chapter 4, the simulation results and experimental results will show much better consistency.

6.4 Concluding remarks

This chapter describes and analyses the experimental results of various parameters of the hybrid pump-controlled asymmetric cylinder drive system, the square wave test is to explore the system performance when in a steady state, the sine wave test is to analyse the system behaviours in dynamics.

Simulation results in square wave and sine wave are compared with experimental measurements, the simulation model does demonstrate correspondence with the overall performance of experimental results, it does pick up the dynamics before and after the velocity reversal. So that the model simulates the hybrid pump-controlled asymmetric cylinder drive system can be utilised for further energy comparison in the next chapter.

Chapter 7 Energy Efficiency Comparison

This hybrid pump-controlled asymmetric cylinder drive system combines the advantages and disadvantages of both valve-controlled and pump-controlled hydraulic applications, its performance has been described and analysed in previous chapters, but its energy efficiency character is not revealed. This chapter utilises the validated simulation model of the hybrid pump-controlled system to calculate its energy efficiency during operation and compare it with that of the valve-controlled system.

7.1 Contrast from theory to practice

Various types of power source are utilised in different hydraulic applications, and their efficiencies are not the same. Besides, as the valve-controlled system does not physically exist, the efficiency of its power source is not measurable. Therefore, the efficiency of the power source is outside the scope of the efficiency calculation in this chapter.

The proposed hybrid pump-controlled asymmetric cylinder drive system is depicted in Fig. 7-1, all the valves in the system are oversized to reduce the throttle losses as much as possible. The needle valve is manually adjusted to reduce the pressure oscillations during operation to improve the system stability performance, but it introduces some throttle losses to the system. So that the hybrid pump-controlled asymmetric cylinder drive system includes the characters of no throttle losses from the pump-controlled only system (energy efficient) and steady performance from the valve-controlled only system (with throttle losses).

This hybrid pump-controlled asymmetric cylinder drive system stays in the middle place of the pump-controlled and valve-controlled system, it should be able to balance the energy efficiency and performance stability. Its performance has been identified in previous chapters but its energy efficiency is unrevealed. One of the best ways to identify its merits is to compare energy efficiency with the valve-controlled system.

However, there is no valve-controlled asymmetric cylinder system available in

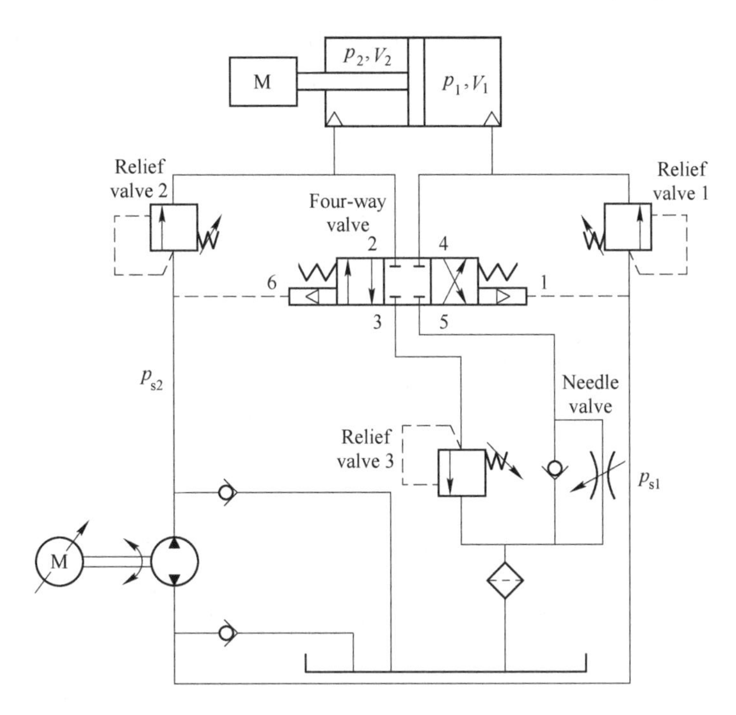

Fig. 7-1 Proposed hybrid pump controlled asymmetric cylinder drive system

realistic, and due to the budget limits, it is difficult to measure the power input of the hybrid pump-controlled system. As both valve-controlled (Moog valve controlled[32]) and hybrid pump-controlled simulation models have been validated in previous chapters, the models can be utilised to calculate and compare their energy efficiencies.

To make two totally different designs comparable, the sizing of the components of both valve-controlled and hybrid pump-controlled asymmetric cylinder system must be the same as each other. The design of the valve-controlled asymmetric cylinder system refers to the Moog valve controlled asymmetric cylinder drive system[32] as in Fig. 7-2 with the proportional control of the position. But open loop configuration is used in this chapter.

The asymmetric cylinders and relevant parameters in both valve-controlled and hybrid pump-controlled system models are set to be the same in this chapter as in Table 7-1.

The energy efficiency is calculated by the ratio of power output to the load and the power generated by the pump, hence:

For the valve-controlled cylinder drive system, the power input to the system can be calculated by Eq. (7-1):

$$p_s \cdot Q_{in} = \text{Power}_{input} \tag{7-1}$$

where p_s is the supply pressure, N/m^2; Q_{in} is the flow rate delivered into the system, m^3/s; $Power_{input}$ is the power input of the system, W.

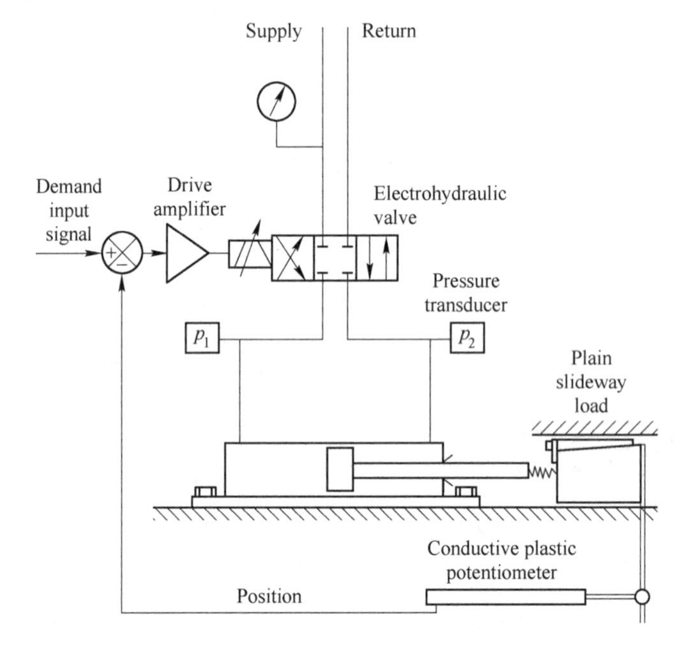

Fig. 7-2 Moog valve controlled asymmetric cylinder system[32]

Table 7-1 Parameters in both valve and hybrid pump controlled system simulation

Cylinder	Piston area A_1	Rod side area A_2	Stroke
	8.04 cm^2	5.50 cm^2	300 mm
Gravity load	20 kg (include tray and piston rod)		
Supply pressure	7 MPa (valve controlled system only)		

The flow rate Q_{in} delivered to the valve-controlled system is calculated by different equations for cylinder extending and retracting, during cylinder extending, the Q_{in} is the flow rate delivered to the piston side chamber, so that it is revealed by:

$$C_D \sqrt{\frac{2}{\rho}} A_v \sqrt{p_s - p_1} = Q_{in} \tag{7-2}$$

During the cylinder retracting, the Q_{in} is the flow rate delivered into the rod side chamber, it revealed by:

$$C_D \sqrt{\frac{2}{\rho}} A_v \sqrt{p_s - p_2} = Q_{in} \tag{7-3}$$

The instantaneous power output of the valve-controlled system in the extending and retracting is calculated by Eq. (7-4):

$$v \cdot p_{LE} \cdot A_E = \text{Power}_{output} \tag{7-4}$$

where v is the velocity of the asymmetric cylinder, m/s; p_{LE} is the effective load pressure, N/m^2; A_E is the effective area of the cylinder, m^2.

For the hybrid pump-controlled asymmetric cylinder drive system, the power input to the system is revealed by Eq. (7-5) and Eq. (7-6):

When cylinder in the extending state, the input power is calculated by:

$$p_{s1} \cdot Q_p = \text{Power}_{input} \tag{7-5}$$

When cylinder in the retracting state, the input power is calculated by:

$$p_{s2} \cdot Q_p = \text{Power}_{input} \tag{7-6}$$

where p_{s1} is the charging pressure in the pipeline when cylinder extending, N/m^2; p_{s2} is the charging pressure in the pipeline when cylinder retracting, N/m^2; the Q_p is the flow rate generate by pump, m^3/s.

The power output of the hybrid pump-controlled system is calculated by Eq. (7-7).

When the cylinder in the extending and retracting state, the instantaneous output power is:

$$v(p_1 A_1 - p_2 A_2) = \text{Power}_{output} \tag{7-7}$$

where p_1 and p_2 are independently simulated, N/m^2, and they indicate piston side chamber pressure and rod side chamber pressure respectively.

Eq. (7-4) and Eq. (7-7) are equal as:

$$v(p_1 A_1 - p_2 A_2) = v \cdot p_{LE} \cdot A_E = \text{Power}_{output} \tag{7-8}$$

The p_{LE} is effective load pressure that can be directly obtained from valve-controlled system model simulation, while chamber pressures p_1 and p_2 are directly obtained from the hybrid pump-controlled system model simulation.

7.2 Undercurrent energy efficiency contest

To compare the energy efficiency between the valve-controlled and the hybrid pump-controlled asymmetric cylinder drive system, the validated valve-controlled and hybrid pump-controlled simulation models are utilised to calculate the energy efficiency, relevant parameters and calculation methods are revealed in the last section.

The components' sizing of the valve-controlled system is consistent with the hybrid pump-controlled system, and their energy efficiencies are calculated with no-load and loaded conditions. In the no-load condition, the output power should only be affected by friction. In the loaded condition, assuming the asymmetric cylinder is placed vertically and there is a 20 kg gravity load constantly applied on the cylinder, its output power is affected by the gravity load and friction.

7.2.1 ±50 mm/s motion energy simulation without load

This section starts with the energy power efficiency calculation of the valve-controlled asymmetric cylinder drive system without load, a command signal is sent to the system to achieve a ± 50 mm/s square wave motion. The energy input and output power simulation results are described as in Fig. 7-3.

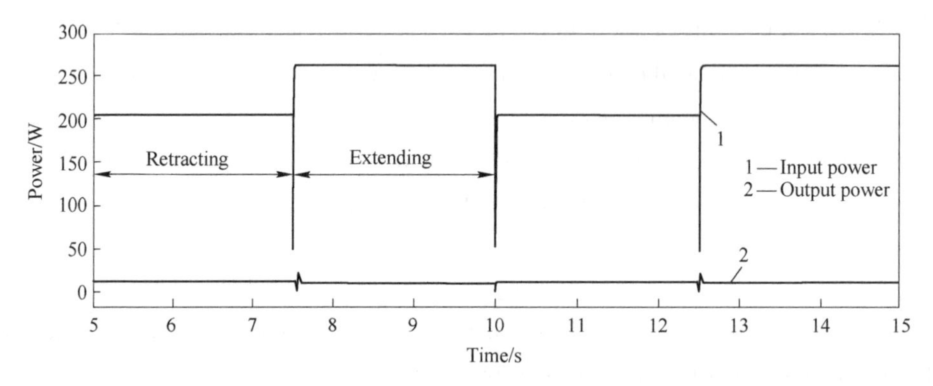

Fig. 7-3 Energy input and output of valve controlled system when cylinder under ±50 mm/s square wave without load

The blue curve is the power generated by the pump and the orange curve depicts the output power applied on the cylinder. It can be observed that the power input during cylinder extending is higher than that during retracting, which is due to the piston side area A_1 is larger the rod side area A_2. In order to implement the ±50 mm/s square wave cylinder motion, a higher flow rate must be delivered to the piston side chamber to achieve the same velocity when cylinder is retracting. As the supply pressure p_s is a constant, the input power during cylinder extending will be larger than that during retracting.

The orange curve output power in Fig. 7-3 is not clear to be observed, the scale difference between input and output power is too large. To make the output power curve analysable, its curve is depicted in Fig. 7-4.

The output power in retracting is larger than that in extending as depicted in Fig. 7-4, this difference is due to the existence of the new friction model. In Chapter 5, pressure

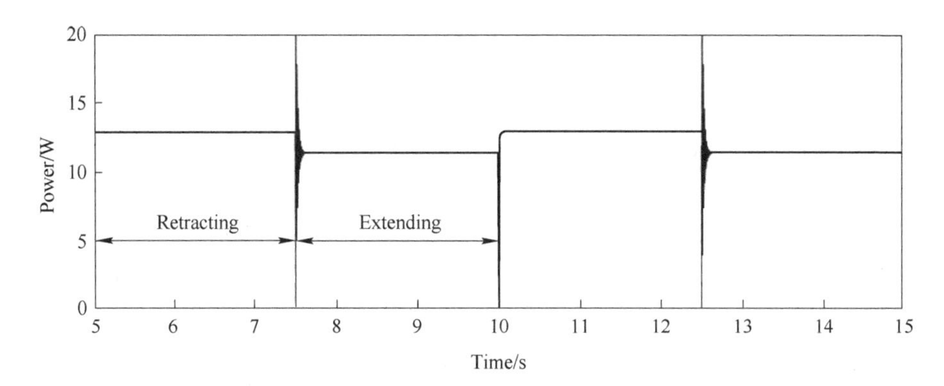

Fig. 7-4 Output power of the valve controlled system under ±50 mm/s square wave without load

difference term is introduced into the new friction model, the pressure difference of the valve-controlled system simulation is depicted in Fig. 7-5.

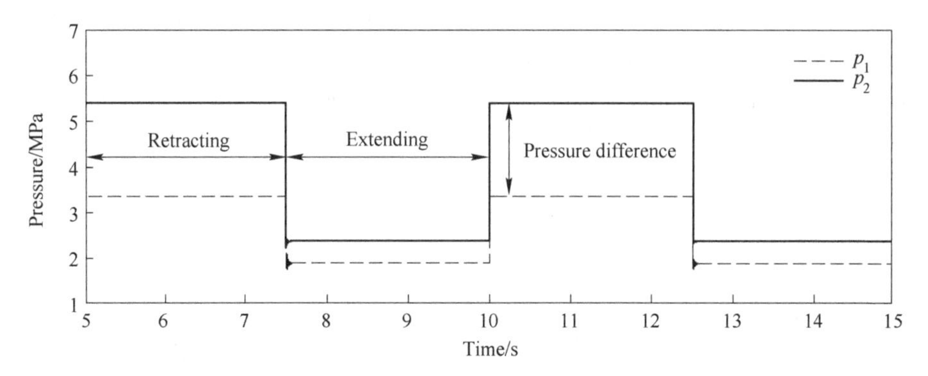

Fig. 7-5 Pressure difference of valve-controlled system simulation under ±50 mm/s square wave without load

It can be observed that the pressure difference in the retracting state is much larger than that in the extending state, leading to a larger friction force when the cylinder in the retracting state. Therefore, the system will output more power to overcome friction in retracting state as in Fig. 7-4.

The input and output power curves in Fig. 7-3 are to analyse the valve-controlled system energy efficiency under steady states. To identify its energy efficiency performance in dynamic, a sine wave signal is sent to the system to achieve the ±50 mm/s sine wave motion on the cylinder, whose input and output power curves are depicted in Fig. 7-6.

The input and output power curves show similar characters to the curves when the cylinder in square wave motion, the peak power input when cylinder in extending state

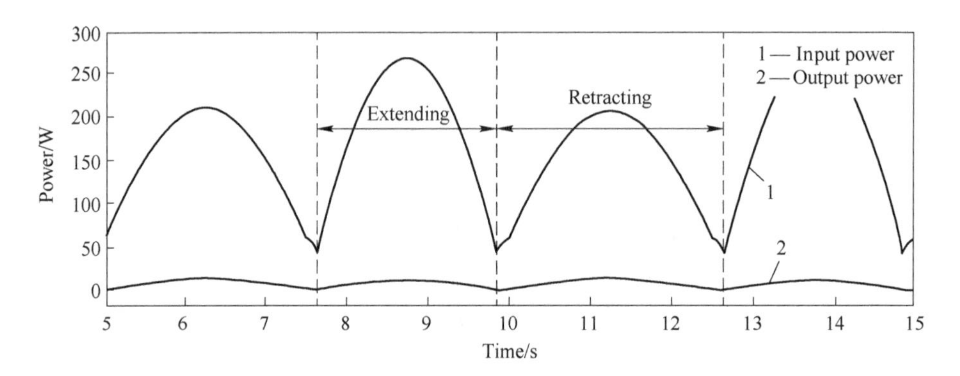

Fig. 7-6 Power input and output when valve controlled system in a sine wave motion under ±50 mm/s square wave without load

is larger than that when the cylinder in retracting state, and this difference can be recognised as an inherent character when a valve-controlled asymmetric cylinder drive system is operating at a symmetric motion.

The detailed output power is still not observable in Fig. 7-6 and it is depicted as in Fig. 7-7.

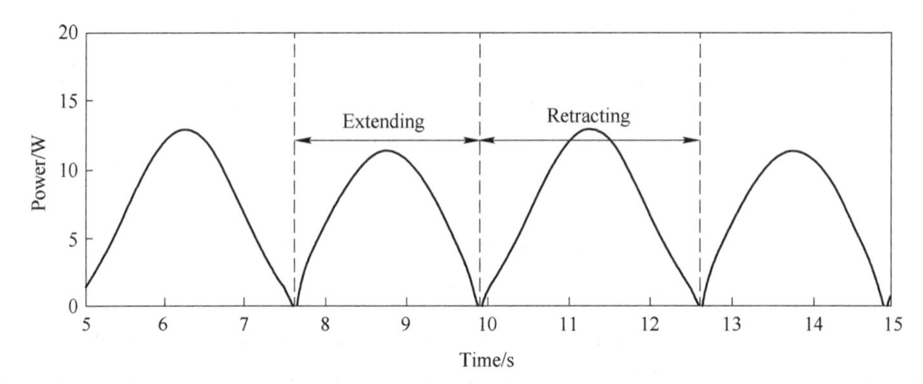

Fig. 7-7 Output power curve when valve controlled system in a sine wave motion under ±50 mm/s square wave without load

The peak output power in the retracting state is larger than that in the extending state as depicted in Fig. 7-7, this phenomenon can be interpreted by the same reason as the square wave simulation in Fig. 7-4. The time duration of retracting state is longer than that of the extending state as depicted in Fig. 7-8, which is caused by the biased sine wave command signal. There are non-smooth parts when the velocity changes its direction, which is caused by the underlap region in the Moog four-way valve in the valve-controlled system. As the valve is partially opened during operation, the underlap region in the cylinder extending state occupies 23% and 16% when the system in the retracting state.

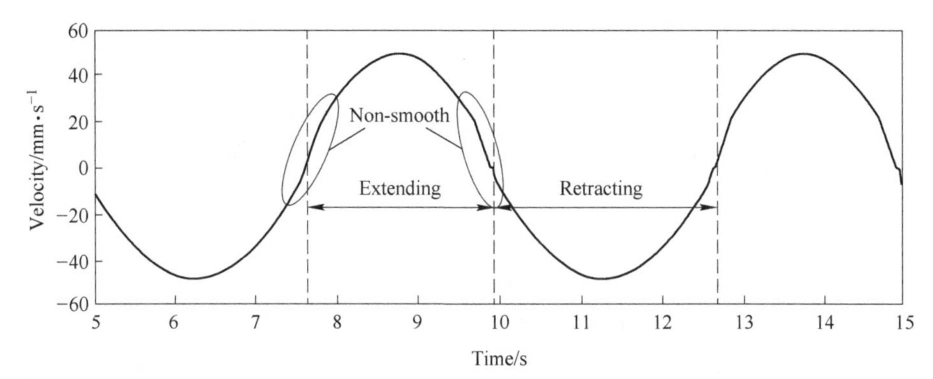

Fig. 7-8 Valve-controlled system cylinder velocity in sine wave motion under ±50 mm/s square wave

For the hybrid pump-controlled asymmetric cylinder drive system, a signal is sent to the system to achieve a ±50 mm/s square wave motion, its input and output power curves are depicted as in Fig. 7-9.

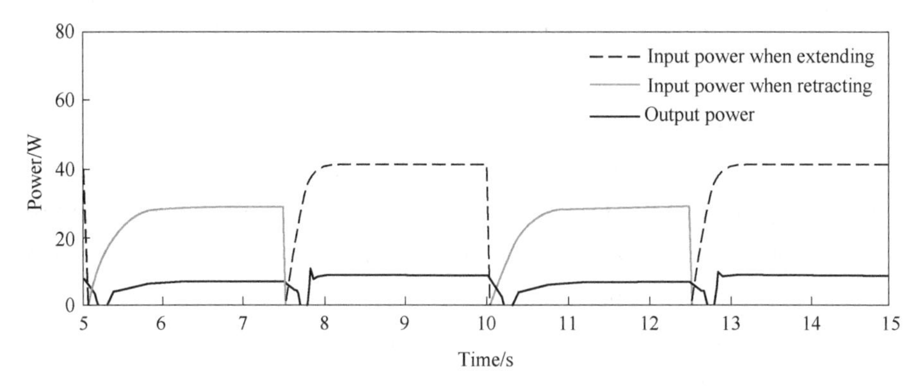

Fig. 7-9 Input and output power of the hybrid pump-controlled system when cylinder in ±50 mm/s square wave motion

The orange curve is the input power when the cylinder is retracting, the blue curve is the input power when the cylinder is extending, the black curve is the output power. Compared with the curves in Fig. 7-3 of the valve-controlled system, the energy efficiency of the hybrid pump-controlled system is much better, but the energy input response time constant of the hybrid pump-controlled system is larger than the valve-controlled one.

The energy power input when the cylinder in the extending state is still higher than that when the cylinder in the retracting state, which can be observed in both Fig. 7-3 and Fig. 7-9. But a difference can be found in the output power curve, when the cylinder in the extending state the output power of hybrid pump-controlled system is larger than that in the retracting state. As there is no load applied, the output power is only utilised to overcome the friction force. The difference is caused by the friction force, the new friction model includes pressure difference square term and the pressure difference in Fig. 6-15 is not as large as that of the valve-controlled system in Fig. 7-5, so that the Coulomb and stiction friction force term in Eq. (5-14) dominate the overall friction force.

To make this argument clear, the steps are given as below to analyse the boundary condition that when the output power of extending and retracting are equal.

(1) As the output power is determined by the friction force, the output power is decided by

$$F_{\mathrm{fss}} = F_{\mathrm{c}} + (F_{\mathrm{s}} - F_{\mathrm{c}})\mathrm{e}^{-\left|\frac{v}{v_{\mathrm{s}}}\right|^{\alpha}} + K(p_1 - p_2)^2 v \tag{7-9}$$

(2) For ±50 mm/s square wave motion, if the extending and retracting friction force is equal, the pressure difference values must satisfy:

$$110 + 21.27 + 2.5 \times 10^{-8} p_{\mathrm{d_{extend}}}^2 \times 0.05 = 80 + 32.5 + 4 \times 10^{-9} p_{\mathrm{d_{retract}}}^2 \times 0.05 \tag{7-10}$$

(3) Rearrange it into

$$9.385 \times 10^{10} + 6.25 p_{\mathrm{d_{extend}}}^2 = p_{\mathrm{d_{retract}}}^2 \tag{7-11}$$

(4) If the pressure differences are not satisfied in above equation, the output power in the extending and retracting will not be equal, either of them can be larger due to the pressure difference values in both conditions.

(5) For instance, substituting the pressure difference in Fig. 7-5 reveals

$$9.385 \times 10^{10} + 6.25(4 \times 10^5)^2 < (20 \times 10^5)^2 \tag{7-12}$$

(6) So that the absolute value of friction force in the valve-controlled system retracting state is larger than that in the extending state, furthermore, the valve-controlled system will consume more energy in the retracting state than that in the extending state. The same algorithm is able to apply to the hybrid pump-controlled system.

For the energy efficiency of the hybrid pump-controlled system in dynamics, a signal is sent to the system to achieve a ±50 mm/s sine wave motion, its input and output power performance curves are described as in Fig. 7-10.

A similar situation in Fig. 7-10 is found when compared with Fig. 7-6, more input power is required to achieve the same velocity in cylinder extending state than that when

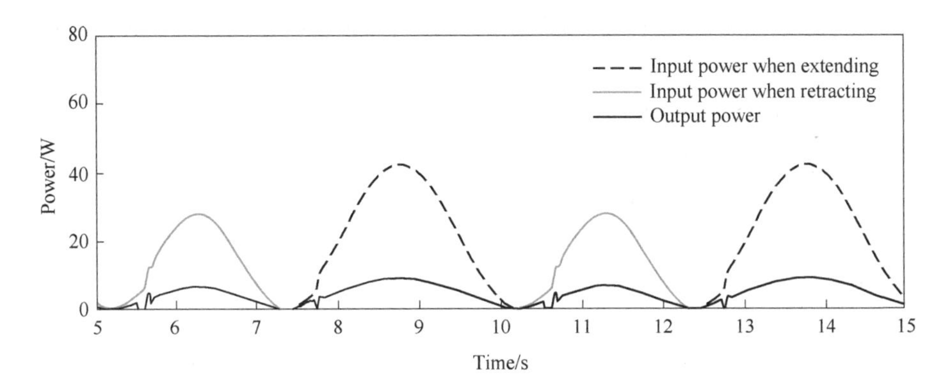

Fig. 7-10 Input and output power when the hybrid pump-controlled system in a ±50 mm/s sine wave motion without load

cylinder is retracting. Its energy efficiency is no doubt much better than the valve-controlled asymmetric cylinder drive system.

The results in above Figures are simulated with ±50 mm/s square wave and sine wave cylinder motion operation without any load. This velocity is relatively slow, and the energy efficiency of the valve-controlled system is very poor compared to the hybrid pump-controlled system at this speed, its energy efficiency may improve with higher speed. Hence, a ±100 mm/s simulation test is carried out.

7.2.2 ±100 mm/s motion energy simulation without load

For the valve-controlled system, its energy efficiency under ±100 mm/s square wave motion without load is depicted in Fig. 7-11.

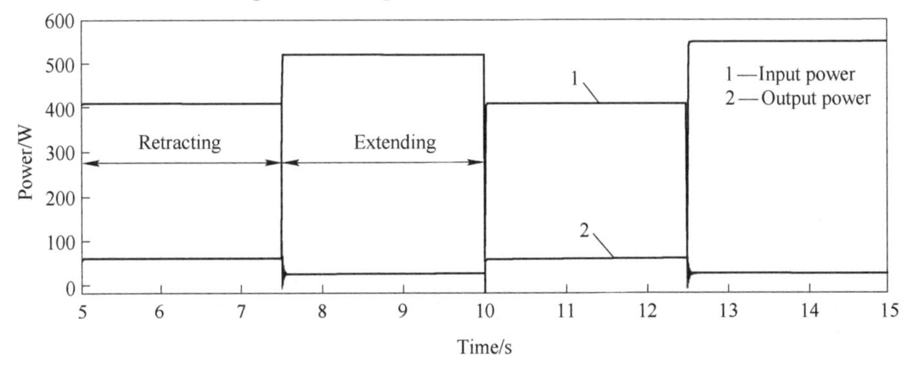

Fig. 7-11 Input and output power of valve-controlled system under cylinder ±100 mm/s square wave motion without load

The shape of input power is similar to that in Fig. 7-3, but its value is almost doubled. The difference can be explained by Eq. (7-1), in which the supply pressure p_s is a constant, and the double velocity will lead to the double flow rate Q_{in}, thus, the input power is almost doubled.

The output power is depicted as in Fig. 7-12, it is about four times larger in the retracting state compared to the output power in Fig. 7-4. A higher velocity leads to a larger pressure difference in the retracting state, the friction force is amplified by the squared pressure difference term in the new friction model, which leads to the increase of the output power in the retracting state is larger than that in the extending state.

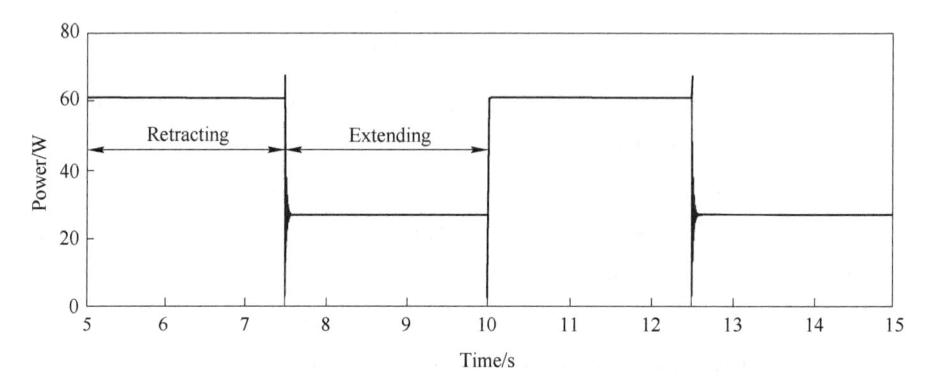

Fig. 7-12 Simulated output power of valve-controlled system when cylinder in ±100 mm/s square wave motion without load

A similar phenomenon can be observed when the valve-controlled asymmetric cylinder drive in ±100 mm/s sine wave motion as in Fig. 7-13 and Fig. 7-14.

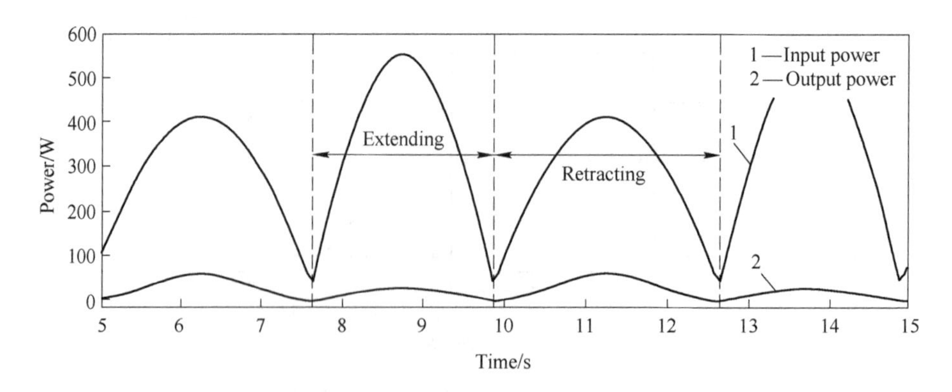

Fig. 7-13 Power input and output of the valve-controlled system when cylinder in ±100 mm/s sine wave motion

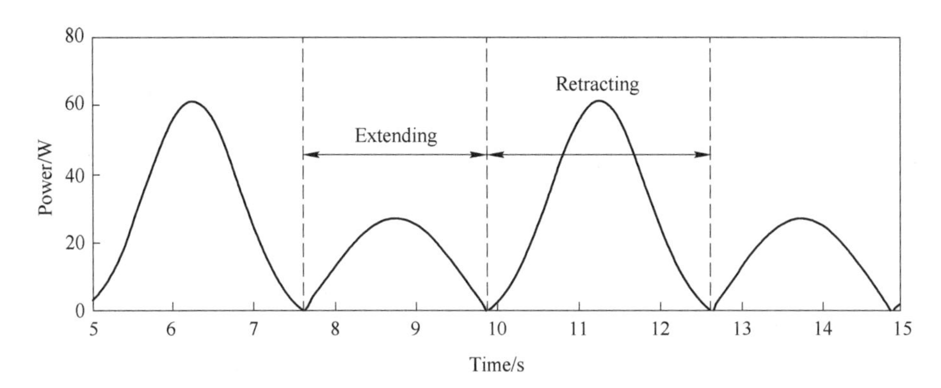

Fig. 7-14 Power output of the valve-controlled system when cylinder in ±100 mm/s sine wave motion without load

For the hybrid pump-controlled asymmetric cylinder drive system, when it is operating at ±100 mm/s square wave motion, its input and output power curves are simulated as in Fig. 7-15.

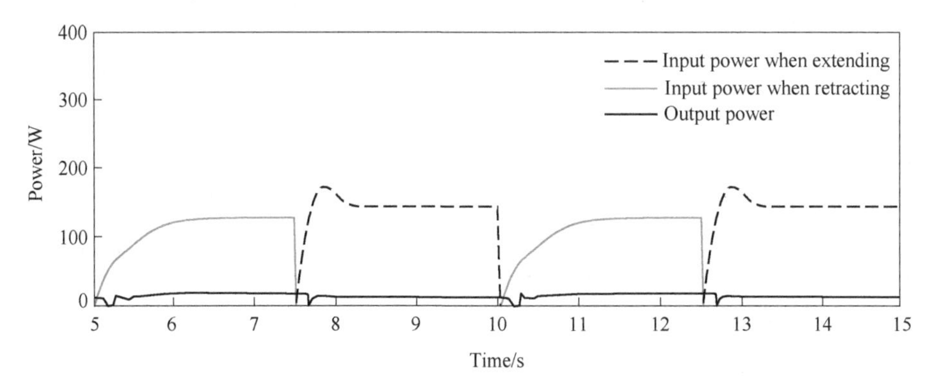

Fig. 7-15 Power input and output of hybrid pump controlled system in ±100 mm/s square wave motion without load

Its energy input and output power when cylinder in ±100 mm/s sine wave motion is revealed in Fig. 7-16.

It can be noticed that the power input in ±100 mm/s motion is larger than that in ±50 mm/s motion (more than double times), this phenomenon is caused by the very small pipe volume in the charging pipeline, and a larger pump flow Q_p will lead to a higher charge pressure.

The energy efficiency is calculated by Output/Input, the power consumption is the area between the power curves and time axis, so the systems energy efficiencies without load are calculated as in Table 7-2.

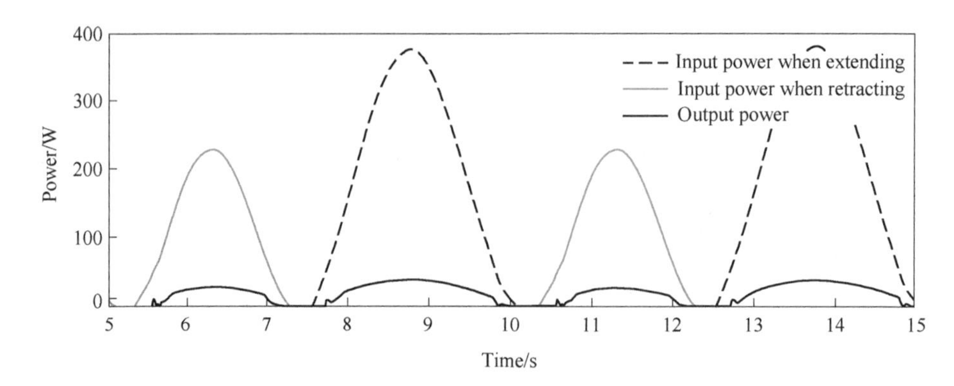

Fig. 7-16 Power input and output of hybrid pump controlled system in ±100 mm/s sine wave motion without load

Table 7-2 Energy efficiency comparison without load

Valve-controlled system energy efficiency	±50 mm/s	Sine wave	3. 70%
		Square wave	4. 16%
	±100 mm/s	Sine wave	7. 39%
		Square wave	8. 59%
Hybrid pump-controlled system energy efficiency	±50 mm/s	Square wave	24. 15%
		Sine wave	24. 65%
	±100 mm/s	Sine wave	13. 97%
		Square wave	14. 60%

Observe the results of the valve-controlled system in Table 7-2, the energy efficiency increased with higher velocity. The valve opening area during operation must be increased to achieve a higher velocity, so that the throttle losses in the valve-controlled system is decreased.

A reversal phenomenon is found in the hybrid pump-controlled system, whose energy efficiency decreases with increased velocity. The efficiency decreasing can be interpreted as that the pump must deliver higher flow rate into the cylinder chambers to increase the cylinder velocity. But due to the relief valve is placed between the charging pipeline and cylinder chamber, a higher flow rate will lead to higher throttle losses. The same situation occurs in the needle valve in the return line.

7. 2. 3 ±50 mm/s motion energy simulation with load supported against gravity

As no-load condition is rare and unrealistic, energy efficiency simulations with 20 kg weight are carried out in this section. This section starts with the energy power efficiency calculation of the valve-controlled asymmetric cylinder drive system with a gravity load, a command signal is sent to the system to achieve a ±50 mm/s square wave motion. The energy input and output power simulation results are described as in Fig. 7-17.

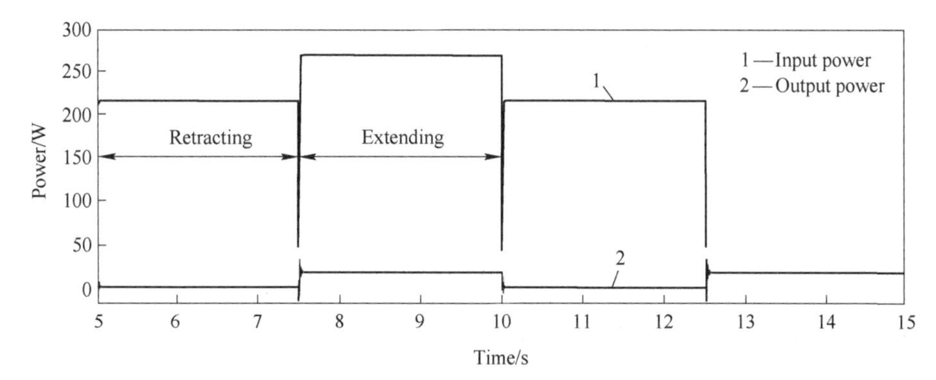

Fig. 7-17 Energy input and output of valve controlled system when cylinder under ±50 mm/s square wave with 20 kg gravity load

The input power of the loaded valve-controlled system is the same as that in the no-load valve-controlled system. But the output power in the extending state is larger than that in the retracting state as in Fig. 7-18.

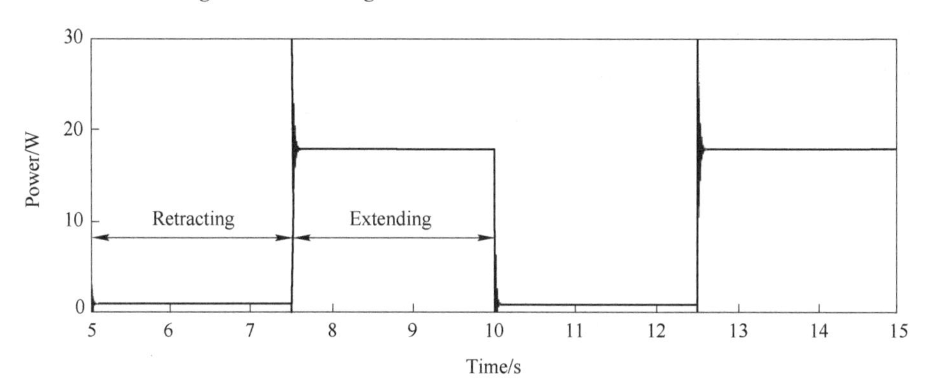

Fig. 7-18 Output power of the valve-controlled system under ±50 mm/s square wave with 20 kg gravity load

This difference is due to the gravity load having a force direction downwards when the cylinder is in the retracting state. Therefore, smaller output power is required to drive the cylinder. When the cylinder is extending, more power is used to overcome the gravity load and friction force. The pressure difference of the valve-controlled system is simulated in Fig. 7-19.

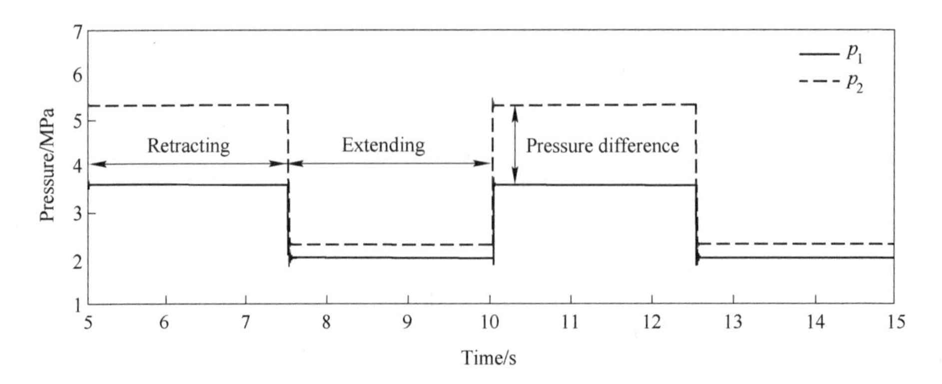

Fig. 7-19 Pressure difference of valve-controlled system simulation under ±50 mm/s square wave with 20 kg gravity load

The pressure difference in above Figure is similar to that in Fig. 7-5, but the chamber pressure p_1 in Fig. 7-19 is larger than that in Fig. 7-5. This outcome is caused by the gravity load on the system. Therefore, the chamber pressure p_1 must provide the necessary force to support the load. When the loaded system is extending, the chamber pressure p_2 in Fig. 7-19 is smaller than that in Fig. 7-5, and this phenomenon is still caused by the gravity load. Therefore, a smaller pressure p_2 is required to balance the force on the cylinder.

To identify the dynamic energy efficiency performance in the valve-controlled system with load supported against gravity, a sine wave signal is sent to the system to achieve the ±50 mm/s sine wave motion on the loaded cylinder. Its input and output power curves are depicted in Fig. 7-20.

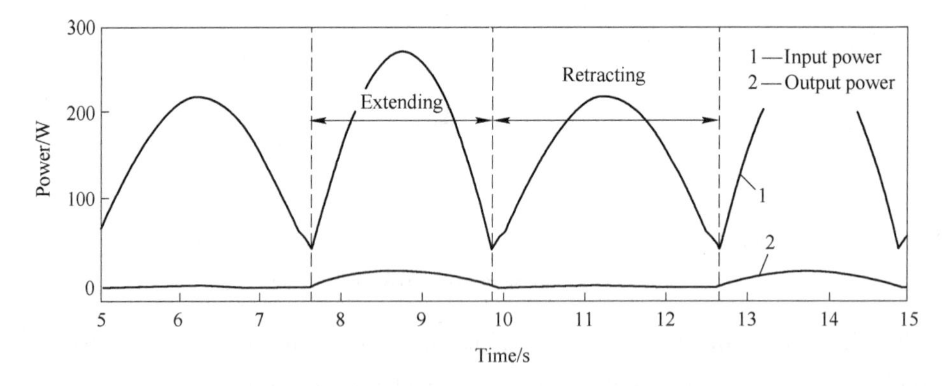

Fig. 7-20 Power input and output when valve controlled system in a sine wave motion under ±50 mm/s square wave with 20 kg gravity load

The input power curve is the same as that in Fig. 7-6, but its output power in the extending is larger than that in the retracting state. The reason is the same as that in the square wave test with load supported against gravity. The detailed output power is still not observable in Fig. 7-20, and the output power is depicted as in Fig. 7-21.

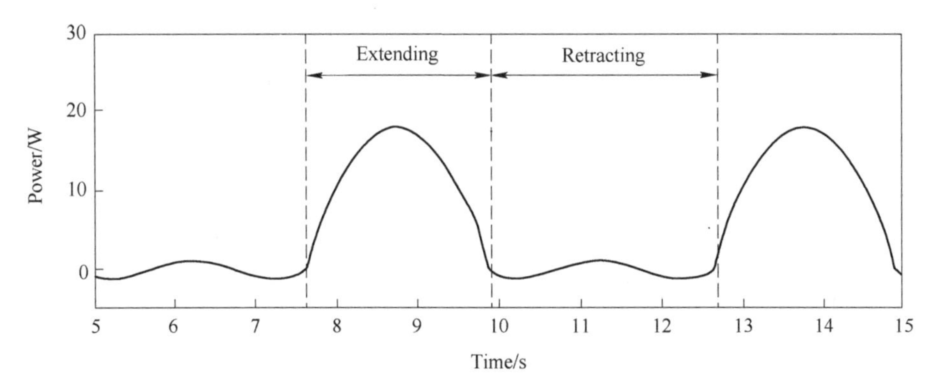

Fig. 7-21 Output power curve when valve controlled system in a sine wave motion under ±50 mm/s square wave with 20 kg gravity load

It can be noticed that a part of the output power curve in the retracting state is below zero, which is caused by the overrunning issue during the load falling in the retracting state.

For the hybrid pump-controlled asymmetric cylinder drive system with load supported against gravity, a signal is sent to the system to achieve a ±50 mm/s square wave motion, its input and output power curves are depicted as in Fig. 7-22.

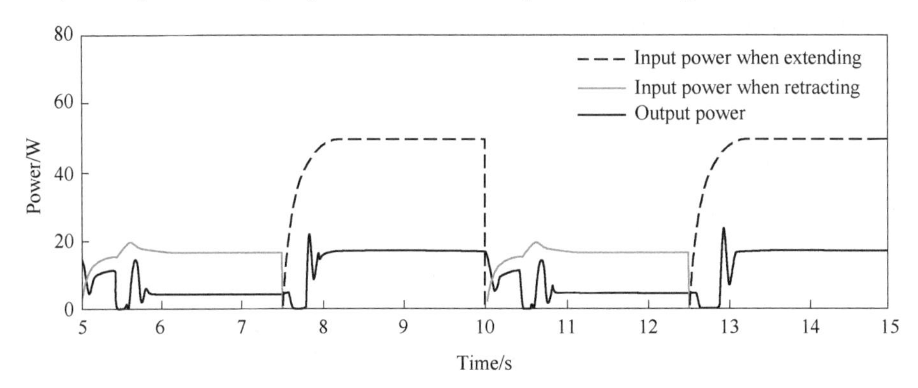

Fig. 7-22 Input and output power of the hybrid pump controlled system when cylinder in ±50 mm/s square wave motion with 20 kg gravity load

Comparing the curves in above Figure with those in Fig. 7-17, the energy efficiency of the hybrid pump-controlled system with load supported against gravity is much better. However, when the hybrid pump-controlled system is retracting, more oscillations can be observed, which is related to the overrunning issue.

A ±50 mm/s sine wave motion is applied on the loaded asymmetric cylinder and the input and output power performance curves are described as in Fig. 7-23.

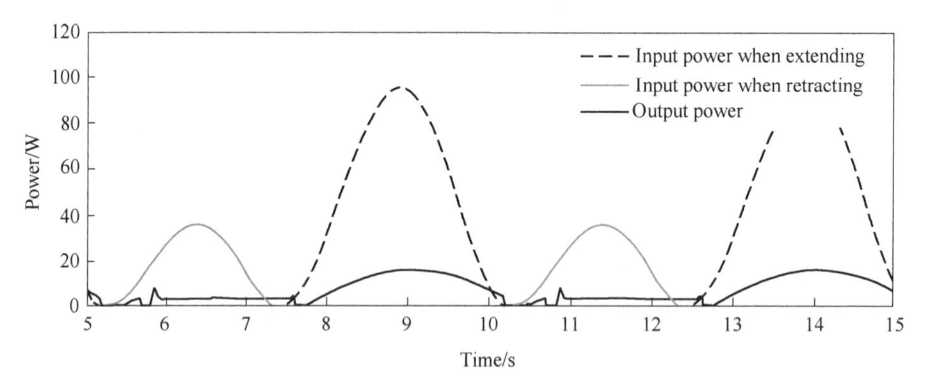

Fig. 7-23 Input and output power when the hybrid pump controlled system in a ±50 mm/s sine wave motion with 20 kg gravity load

Overrunning problem in the retracting state in above figure is observed, which is similar to that in Fig. 7-20.

7.2.4 ±100 mm/s motion energy simulation with load supported against gravity

For the loaded valve-controlled system, the input and output power curves in ±100 mm/s square wave motion are depicted in Fig. 7-24.

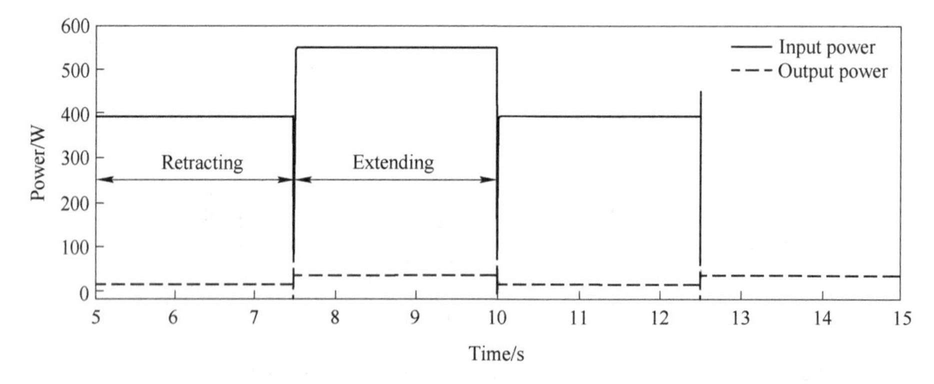

Fig. 7-24 Input and output power of valve controlled system under cylinder ±100 mm/s square wave motion with 20 kg gravity load

The input power of the loaded test in above figure is the same as that in Fig. 7-11. The output power is depicted in Fig. 7-25.

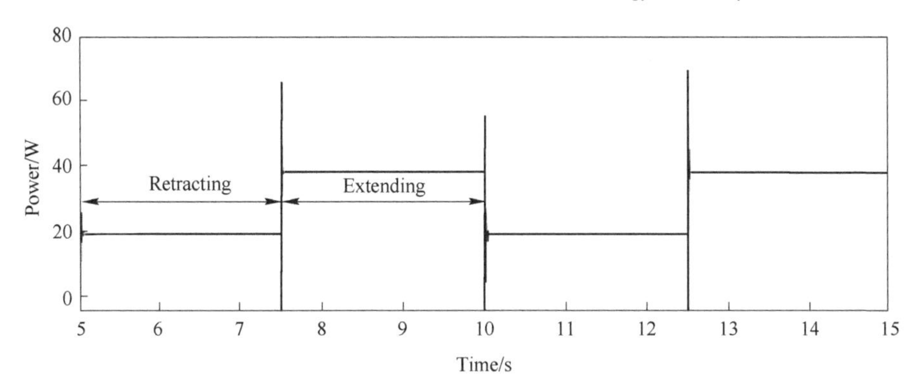

Fig. 7-25 Simulated output power of valve-controlled system when cylinder in ±100 mm/s square wave motion with 20 kg gravity load

The shape of the output power curve in above figure is similar to that in Fig. 7-18, but its overall power output is larger due to the higher velocity of the cylinder.

The input and output power curves of the valve-controlled asymmetric cylinder drive in the ±100 mm/s sine wave motion with load supported against gravity are depicted in Fig. 7-26 and Fig. 7-27.

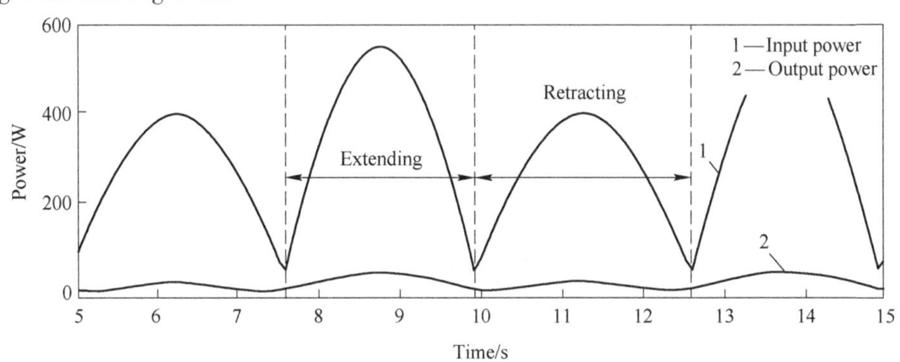

Fig. 7-26 Power input and output of the valve-controlled system when cylinder in ±100 mm/s sine wave motion with 20 kg gravity load

The output power in above Figure is increased due to the doubled velocity, and its behaviour is similar to that in Fig. 7-21, meanwhile, the overrunning issue is still observable.

When the hybrid pump-controlled asymmetric cylinder drive system with load supported against gravity is operating at ±100 mm/s square wave motion, its input and output power curves are simulated as in Fig. 7-28.

The energy input and output power curves when the cylinder in ±100 mm/s sine wave motion are depicted in Fig. 7-29.

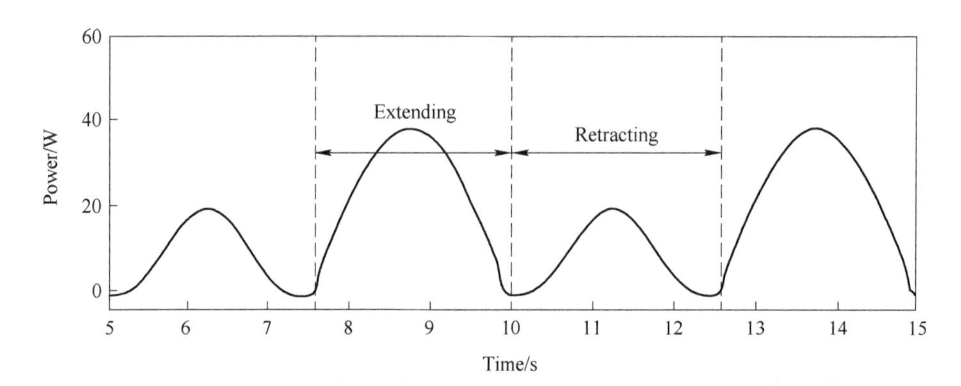

Fig. 7-27 Power output of the valve-controlled system when cylinder in ±100 mm/s sine wave motion with 20 kg gravity load

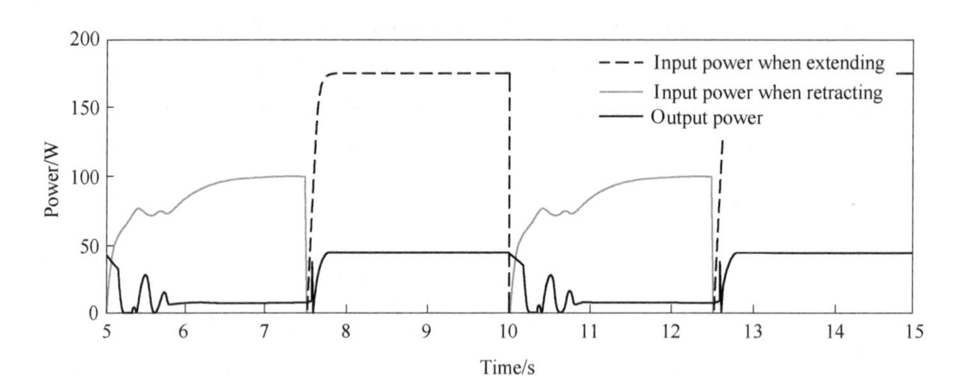

Fig. 7-28 Power input and output of the hybrid pump controlled system in ±100 mm/s square wave motion with 20 kg gravity load

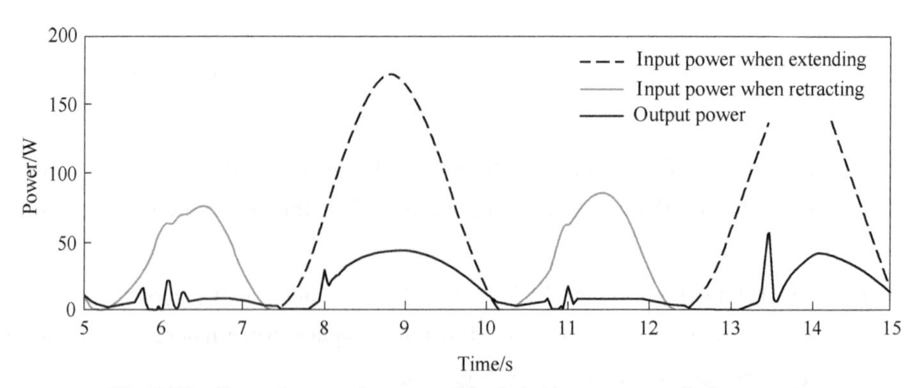

Fig. 7-29 Power input and output of the hybrid pump-controlled system in ±100 mm/s sine wave motion with 20 kg gravity load

Though the output oscillations in the retracting state are observable, the overrunning issue is much better compared to that in Fig. 7-23. As a higher cylinder velocity indicates a higher flow rate passing through the needle valve with a fixed opening area, a higher flow rate will lead to a higher chamber pressure p_1. Therefore, a higher velocity in the retracting state can improve the overrunning problem. The system energy efficiency is calculated as in Table 7-3.

Table 7-3 Energy efficiency comparison with gravity load

Valve-controlled system energy efficiency	±50 mm/s	Sine wave	2.75%
		Square wave	3.29%
	±100 mm/s	Sine wave	4.95%
		Square wave	5.88%
Hybrid pump-controlled system energy efficiency	±50 mm/s	Sine wave	24.65%
		Square wave	24.78%
	±100 mm/s	Sine wave	27.59%
		Square wave	27.14%

Though the gravity load reduces the required output power in the retracting state, the input power of the valve-controlled system remains the same. Therefore, the overall energy efficiency of the valve-controlled system with load supported against gravity is lower compared to the efficiency in the no-load situation.

As overrunning issue happened in the ±50 mm/s motion simulation of the loaded hybrid pump-controlled system, the efficiency calculation will be inaccurate. So that the analysis and comparison of the efficiency are only based on the ±100 mm/s simulation data.

The overall efficiency of the loaded hybrid pump-controlled system in the ±100 mm/s motion is higher than the efficiency of the no-load system. The phenomenon is also caused by the gravity load, which reduces the required output power in the retracting state. As the pump input power in this hybrid pump-controlled system can adapt itself to the output power requirement, a part of the input power can also be saved.

The energy efficiency performance of the hybrid pump-controlled system still has a massive potential to be optimised, the optimised approach can be summarised as:

(1) Replacing the RBAE relief valve 1 and 2 with oversized unloading valves or solenoid valves, to reduce their throttle losses in operation.

(2) Replacing the DDDC pilot shifted four-way valve with a solenoid valve, so that

the relief valve 1 and 2 in Fig. 7-1 can be removed and the throttle losses in both relief valves can be avoided.

(3) For stability purpose, the manually adjusted needle valve can be optimised with different opening areas under different loading and cylinder velocity conditions.

Due to the budget limits, the optimised objectives are not applied in this research, but its performance of the hybrid pump-controlled concept is evaluated and assessed.

7.3 Conclusion remarks

The first section of this chapter proposed a methodology to calculate the energy efficiency of the valve-controlled and hybrid pump-controlled asymmetric cylinder drive systems. As simulation models are validated by experimental results already, the validate models are utilised to generate the energy power performance curves.

To make both models comparable, the component sizing parameters in the valve-controlled system are modified to be the same as those in the hybrid pump-controlled system. These results indicate that the energy power efficiency of the hybrid pump-controlled system is no doubt much better than the valve-controlled system.

The better efficiency is mainly caused by the power input required by the hybrid pump-controlled system is much less than the valve-controlled system. Though the valve-controlled system can improve its energy efficiency by reducing its supply pressure p_s, the design structure of the hybrid pump-controlled decides that its energy efficiency will always be better than a conventional valve-controlled system.

The four-way valve in this hybrid pump-controlled system operates the same as Can Do's variable supply pressure valve-controlled system[86], in which the valve only operates in its maximum opening area, and her system achieved 70% improvement in the energy efficiency aspect. But her design circuit is complicated and the usage of servo valves increases the cost, the hybrid pump-controlled system in this research can achieve similar energy efficiency improvement with much lower cost and simpler design.

Chapter 8 Research Ends, Future Road, Innovation Continues

The work that is done in this thesis addresses three major aims: improve the modeling of asymmetric cylinder drive system; improve the driving of asymmetric cylinder system at low speed and velocity reversal; and combine the advantages of valve-controlled and pump-controlled asymmetric cylinder drive system for energy efficiency purpose. The findings are highlighted and further works are given in this chapter.

8.1 Conclusions

Several significant achievements are obtained from works done in this thesis, including the improvement of model simulation time cost, the implementation of a new friction model and implementation of a hybrid pump controlled asymmetric cylinder drive system.

8.1.1 Improve the simulation model

The component linking model for a Moog valve-controlled system is originally constructed in Fortran, it validated here in Matlab Simulink. A further nonlinear research in the underlap is carried out based on a validated simulation model, an analytical solution when the valve in the underlap region is implemented to replace the original numerical solution.

(1) The Moog valve-controlled system simulation results in Matlab Simulink are validated against published results in reference [63].

(2) The chamber pressure behaviour when the valve in the underlap region is revealed by simulation model in Simulink.

(3) X factor is utilised for the generalised concept equations and used in "component linking" method modeling. An analytical solution for X factor when the valve

in the underlap region is implemented.

(4) For a certain value of area ratio γ, the value of X factor is only affected by the travel percentage.

(5) The X factor roots distribution is revealed by an equation generated by Matlab CFTOOL.

(6) The analytical solution is approximately 200 times more computationally efficient than the original numerical solution method.

8.1.2 New friction model

The most suitable friction model for hydraulic application is the LuGre model, but the model only includes the velocity term. Other factor influences are investigated, and a new friction model is implemented on current LuGre model. The simulation results of new friction model are compared with experimental results in various loads condition.

(1) The new friction model produced here introduces additional pressure difference term and acceleration term in addition to the velocity term of the original LuGre model.

(2) The new friction model shows much better consistency with experimental results than the original LuGre model.

(3) Parameters for the hybrid pump-controlled system are identified with various tests.

(4) Different inertia load conditions only affect the Coulomb friction F_c and stiction friction F_s in the new friction model.

8.1.3 Hybrid pump-controlled asymmetric cylinder drive system

For the energy efficiency purpose, a hybrid pump-controlled system that combines the advantages of the valve-controlled and pump-controlled system is implemented. An oversized pilot-shifted four-way valve is selected to reduce the throttle losses and for load holding purpose. Square and sine wave experimental tests are carried out, simulation results are compared. Its system performance and energy efficiency are analysed.

(1) The hybrid pump-controlled system is constructed at a low cost, all its operation is completed passively.

(2) The stall motion when cylinder velocity at low and reversal is caused by valves operation and friction force.

(3) The simulation model demonstrates consistency with the overall performance of

the test rig. The experimental dynamics when cylinder velocity is low and during reversal is captured by the model.

(4) The energy efficiency of the hybrid pump controlled asymmetric cylinder drive is much better than that of the valve-controlled system.

(5) The output power of the hybrid pump-controlled and the valve-controlled system is dominated by the friction force with no load applied.

(6) The gravity load can reduce power input and output of the hybrid pump-controlled system.

(7) Under symmetric cylinder velocity motion, the friction force is not equal when cylinder in the extending or in the retracting state, which is caused by the pressure difference term in the new friction model.

8. 2 New chapter in the future

The works presented in this thesis can keep improving and further research not limiting to:

(1) *X* factor roots analytical solution when valve in underlap region, the roots curves combination is the same as the numerically solved one, some mathematical theory must be able to explain this phenomenon.

(2) When the valve in the underlap region, the performance of the valve-controlled asymmetric cylinder drive system is modeled by simulation, but the experimental validation needs to be carried out.

(3) The relief valve 1 and 2 can be replaced with two oversized unloading valves or solenoid valves, so that the throttle losses during operation can be avoided.

(4) For the hybrid pump controlled asymmetric cylinder drive system, the pilot-shifted four-way valve is operated purely mechanically, a solenoid four-way valve can be used to replace it and the relief valve 1 and 2 can be removed. So that the system is able to offer better dynamic performance, this should be able to solve the stall motion problem.

(5) The manually adjusted needle valve opening area during operation may be optimised for a better balance between damping and throttle losses, this optimisation can be done in simulation first and validate in the experimental test.

(6) A close loop controller can be added to the system, the controller adjusts the servomotor to control the flow rate and chamber pressures to have a better system

dynamic response.

(7) Replace the manually adjusted needle valve with a servo type one. The servo needle valve opening is adjusted due to the cylinder velocity by the controller.

(8) The new friction model can be further explored to see if the parameters are affected by different load and oil temperature.

References

[1] HE X, WANG X, FENG Z, et al. Nonlinear modeling of electrohydraulic servo injection molding machine including asymmetric cylinder [C]//Proceedings of the 2003 American Control Conference. United States: IEEE, 2003, 4: 3055-3059.

[2] KELLY A L, WOODHEAD M, COATES P D. Comparison of injection molding machine performance [J]. Polymer Engineering & Science, 2005, 45 (6): 857-865.

[3] POOMPHOCHANA K, PRATUMSUWAN P, PO-NGAEN W. Energy saving in electro-hydraulic system using impedance sensing [C]//3rd International Conference on Electric and Electronics. [S. I.]: Atlantis Press, 2013: 245-248.

[4] QI H, LIU Z, LANG Y. Symmetrical valve controlled asymmetrical cylinder based on wavelet neural network [J]. Engineering Computations, 2017, 34 (7): 2154-2167.

[5] RUDERMAN M. Full-and reduced-order model of hydraulic cylinder for motion control [C]// IECON 2017-43rd Annual Conference of the IEEE Industrial Electronics Society. United States: IEEE, 2017: 7275-7280.

[6] VIERSMA T J, ANDERSEN B W. Analysis, synthesis, and design of hydraulic servosystems and pipelines [J]. Journal of Dynamic Systems Measurement and Control Transactions of the Asme, 1981 (103): 73.

[7] OWEN W S, CROFT E A. The reduction of stick-slip friction in hydraulic actuators [J]. IEEE/ ASME Transactions on Mechatronics, 2003, 8 (3): 362-371.

[8] AL-GHATHIAN F M M, TARAWNEH M S. Friction forces in O-ring sealing [J]. American Journal of Applied Sciences, 2005, 2 (3): 626-632.

[9] ZHANG W, LI G, WANG L. Application of the improved variable arguments PID in asymmetric hydraulic cylinder electro-hydraulic servo system [C]//2015 IEEE Fifth International Conference on Big Data and Cloud Computing. United States: IEEE, 2015: 223-226.

[10] LOVE L J, LANKE E, ALLES P. Estimating the impact (energy, emissions and economics) of the US fluid power industry [R]. Oak Ridge: Oak Ridge National Laboratory, TN, 2012.

[11] MANRING L H, MANRING N D. Mapping the efficiency of a double acting, single-rod hydraulic-actuator using a critically centered four-way spool valve and a load-sensing pump [J]. Journal of Dynamic Systems, Measurement, and Control, 2018, 140 (9): 091017.

[12] WANG T, WANG Q. Efficiency analysis and evaluation of energy-saving pressure-compensated circuit for hybrid hydraulic excavator [J]. Automation in Construction, 2014, 47: 62-68.

[13] WANG T, WANG Q, Lin T. Improvement of boom control performance for hybrid hydraulic excavator with potential energy recovery [J]. Automation in Construction, 2013, 30: 161-169.

[14] MAN Z, DING F, DING C, et al. Study of an energy regeneration system with accumulator for hydraulic impulse testing equipment [J]. Strojniški Vestnik-Journal of Mechanical Engineering, 2015, 61 (3): 196-206.

[15] HUOVA M, LINJAMA M, HUHTALA K. Energy efficiency of digital hydraulic valve control systems [R]. SAE Technical Paper, 2013.

[16] XIAO Q, WANG Q, ZHANG Y. Control strategies of power system in hybrid hydraulic excavator [J]. Automation in Construction, 2008, 17 (4): 361-367.

[17] LIN T, WANG Q, HU B, et al. Development of hybrid powered hydraulic construction machinery [J]. Automation in construction, 2010, 19 (1): 11-19.

[18] HAO Y, QUAN L, CHENG H, et al. Potential energy directly conversion and utilization methods used for heavy duty lifting machinery [J]. Energy, 2018, 155: 242-251.

[19] QUAN Z, QUAN L, ZHANG J. Review of energy efficient direct pump controlled cylinder electro-hydraulic technology [J]. Renewable and Sustainable Energy Reviews, 2014, 35: 336-346.

[20] AXIN M, ERIKSSON B, KRUS P. Flow versus pressure control of pumps in mobile hydraulic systems [J]. Proceedings of the Institution of Mechanical Engineers, Part I: Journal of Systems and Control Engineering, 2014, 228 (4): 245-256.

[21] WU H W, LEE C B. Self-tuning adaptive speed control of a pump/inverter-controlled hydraulic motor system [J]. Proceedings of the Institution of Mechanical Engineers, Part I: Journal of Systems and Control Engineering, 1995, 209 (2): 101-114.

[22] TAO J, WANG X, YANG L, et al. Nonovershooting position control for unidirectional proportional pump controlled asymmetric cylinder with proportional controller [C]//2015 International Conference on Fluid Power and Mechatronics (FPM). United States: IEEE, 2015: 866-872.

[23] CHIANG M H, YEH Y P, YANG F L, et al. Integrated control of clamping force and energy-saving in hydraulic injection moulding machines using decoupling fuzzy sliding-mode control [J]. The International Journal of Advanced Manufacturing Technology, 2005, 27: 53-62.

[24] RAHMFELD R, IVANTYSYNOVA M. Energy saving hydraulic displacement controlled linear actuators in industry applications and mobile machine systems [C]//The Fourth International Symposium on Linear Drives for Industry Applications (LDIA03). Birmingham, UK: [S. N.], 2003: 37-40.

[25] WANG L, BOOK W J, Huggins J D. A hydraulic circuit for single rod cylinders [J]. Journal of Dynamic Systems, Measurement, and Control, 2012, 134 (1): 011019.

[26] WANG L, BOOK W J. Using leakage to stabilize a hydraulic circuit for pump controlled actuators [J]. Journal of Dynamic Systems, Measurement, and Control, 2013, 135 (6): 061007.

[27] WILLIAMSON C, IVANTYSYNOVA M. Pump mode prediction for four-quadrant velocity control of valueless hydraulic actuators [C] //Proceedings of the JFPS International Symposium on Fluid Power. Japan: The Japan Fluid Power System Society, 2008: 323-328.

[28] HO T H, AHN K K. Modeling and simulation of hydrostatic transmission system with energy regeneration using hydraulic accumulator [J]. Journal of Mechanical Science and Technology,

2010, 24: 1163-1175.

[29] LI W, WU B, CAO B. Control strategy of a novel energy recovery system for parallel hybrid hydraulic excavator [J]. Advances in Mechanical Engineering, 2015, 7 (10): 1-9.

[30] ZHAO P Y, CHEN Y L, ZHOU H. Simulation analysis of potential energy recovery system of hydraulic hybrid excavator [J]. International Journal of Precision Engineering and Manufacturing, 2017, 18: 1575-1589.

[31] JALAYERI E, IMAM A, SEPEHRI N. A throttle-less single rod hydraulic cylinder positioning system for switching loads [J]. Case Studies in Mechanical Systems and Signal Processing, 2015, 1: 27-31.

[32] LEANEY P G. Component oriented modeling of a valve controlled asymmetric cylinder drive using a generalized formulation for model linking [J]. Proceedings of the Institution of Mechanical Engineers, Part Ⅰ: Journal of Systems and Control Engineering, 1991, 205 (4): 285-298.

[33] TAO J, WANG X, XIONG Z, et al. Modeling and simulation of unidirectional proportional pump-controlled asymmetric cylinder position control system with model predictive control algorithm [C]//2016 IEEE International Conference on Aircraft Utility Systems (AUS). United States: IEEE, 2016: 408-413.

[34] MATTILA J, VIRVALO T. Energy-efficient motion control of a hydraulic manipulator [C]// Proceedings 2000 ICRA. Millennium Conference. IEEE International Conference on Robotics and Automation. Symposia Proceedings. United States: IEEE, 2000, 3: 3000-3006.

[35] DAHER N A, IVANTYSYNOVA M. Pump Controlled Steer-by-Wire System [R]. SAE Technical Paper, 2013.

[36] HEYBROEK K, PALMBERG J O. Applied control strategies for a pump controlled open circuit solution [J]. Digitala Vetenskapliga Arkivet [S. I.], 2008: 39-52.

[37] IMAM A, RAFIQ M, JALAYERI E, et al. Design, implementation and evaluation of a pump-controlled circuit for single rod actuators [C]//Actuators. [S. I.]: MDPI, 2017, 6 (1): 10.

[38] QUAN L. Current state, problems and the innovative solution of electro-hydraulic technology of pump controlled cylinder [J]. Journal of Mechanical Engineering, 2008, 44 (11): 87-92.

[39] HVOLDAL M, OLESEN C. Friction modeling and parameter estimation for hydraulic asymmetrical cylinders [D]. Aalburg: Aalburg University, 2011.

[40] AL-BENDER F. Fundamentals of friction modeling [C]//ASPE Spring Topical Meeting on Control of Precision Systems. United States: ASPE-The American Society of Precision Engineering, 2010, 48: 117-122.

[41] LIU L L, WU Z Y. A new identification method of the Stribeck friction model based on limit cycles [J]. Proceedings of the Institution of Mechanical Engineers, Part C: Journal of Mechanical Engineering Science, 2014, 228 (15): 2678-2683.

[42] MÁRTON L, LANTOS B. Identification and model-based compensation of striebeck friction

[J]. Acta Polytechnica Hungarica, 2006, 3 (3): 45-58.

[43] ARMSTRONG-HÉLOUVRY B, DUPONT P, DE WIT C C. A survey of models, analysis tools and compensation methods for the control of machines with friction [J]. Automatica, 1994, 30 (7): 1083-1138.

[44] RABINOWICZ E. The intrinsic variables affecting the stick-slip process [J]. Proceedings of the Physical Society, 1958, 71 (4): 668.

[45] HESS D P, SOOM A. Friction at a lubricated line contact operating at oscillating sliding velocities [J]. Journal of Tribology, 1990, 112 (1): 147-152.

[46] AL-BENDER F, SWEVERS J. Characterization of friction force dynamics [J]. IEEE Control Systems Magazine, 2008, 28 (6): 64-81.

[47] NACHANE R P, HUSSAIN G F S, IYER K R. Theory of stick-slip effect in friction [J]. CSIR-NIScPR, 1998, 23 (4): 201-208.

[48] LIU Y F, LI J, ZHANG Z M, et al. Experimental comparison of five friction models on the same test-bed of the micro stick-slip motion system [J]. Mechanical Sciences, 2015, 6 (1): 15-28.

[49] BENGISU M T, AKAY A. Stick-slip oscillations: Dynamics of friction and surface roughness [J]. The Journal of the Acoustical Society of America, 1999, 105 (1): 194-205.

[50] AL-BENDER F. On the modeling of the dynamic characteristics of aerostatic bearing films: From stability analysis to active compensation [J]. Precision Engineering, 2009, 33 (2): 117-126.

[51] BONSIGNORE A, FERRETTI G, MAGNANI G. Analytical formulation of the classical friction model for motion analysis and simulation [J]. Mathematical and Computer Modeling of Dynamical Systems, 1999, 5 (1): 43-54.

[52] KARNOPP D. Computer simulation of stick-slip friction in mechanical dynamic systems [J]. Journal of Dynamic Systems Measurement and Control, 1985, 107 (1): 100-103.

[53] TUSTIN A. The effects of backlash and of speed-dependent friction on the stability of closed-cycle control systems [J]. Journal of the Institution of Electrical Engineers-Part Ⅱ A: Automatic Regulators and Servo Mechanisms, 1947, 94 (1): 143-151.

[54] BO L C, PAVELESCU D. The friction-speed relation and its influence on the critical velocity of stick-slip motion [J]. Wear, 1982, 82 (3): 277-289.

[55] DAHL P R. A solid friction model [J]. The Aerospace Corporation, 1968, 18 (1): 1-24.

[56] DAHL P R. Measurement of solid friction parameters of ball bearings [C] //Proceedings of the 6th Annual Symposium on Incremental Motion, Control Systems and Devices. United States: University of Illinois, 1977: 49-60.

[57] HAESSIG D A , FRIEDLAND B. On the modeling and simulation of friction [C] //American Control Conference. United States: IEEE, 1990: 1256-1261.

[58] HAESSIG Jr D A, FRIEDLAND B. On the modeling and simulation of friction [J]. IEEE Xplore, 1991, 40 (3): 419-425.

[59] JOHANASTROM K, CANUDAS-DE-WIT C. Revisiting the LuGre friction model [J]. IEEE Control Systems Magazine, 2008, 28 (6): 101-114.

[60] AL-BENDER F, LAMPAERT V, SWEVERS J. The generalized Maxwell-slip model: a novel model for friction simulation and compensation [J]. IEEE Transactions on Automatic Control, 2005, 50 (11): 1883-1887.

[61] SWEVERS J, AL-BENDER F, GANSEMAN C G, et al. An integrated friction model structure with improved presliding behavior for accurate friction compensation [J]. IEEE Transactions on Automatic Control, 2000, 45 (4): 675-686.

[62] LAMPAERT V, SWEVERS J, AL-BENDER F. Modification of the Leuven integrated friction model structure [J]. IEEE Transactions on Automatic Control, 2002, 47 (4): 683-687.

[63] LEANEY P G. The modeling and computer aided design of hydraulic servosystems [D]. Loughborough: Loughborough University, 1986.

[64] MOOG. 76 series valves catalogue [EB/OL]. 1967. [2019-05-01]. http: //www. moog. com/literature/ICD/76seriesvalves. pdf.

[65] DORMAND J R, PRINCE P J. A family of embedded Runge-Kutta formulae [J]. Journal of Computational and Applied Mathematics, 1980, 6 (1): 19-26.

[66] CLEASBY K G, PLUMMER A R. A novel high efficiency electro-hydrostatic flight simulator motion system [M]//Fluid Power and Motion Control (FPMC 2008). Bath: Centre for PTMC, 2008: 437-449.

[67] ZIMMERMAN J D, PELOSI M, WILLIAMSON C A, et al. Energy consumption of an LS excavator hydraulic system [C]//ASME International Mechanical Engineering Congress and Exposition. Seattle: AMSE, 2007, 42983: 117-126.

[68] JOHNSON J. Exploring an alternative pump control method [EB/OL]. 2007. [2019-05-01]. https: //www. hydraulicspneumatics. com/200/TechZone/HydraulicPumpsM/Article/False/63503/TechZone-HydraulicPumpsM.

[69] SEM. MT30R4 series servo motor data sheet [EB/OL]. [2019-05-01]. https: //forum. linuxcnc. org/media/kunena/attachments/21417/MT30R4Extract. pdf.

[70] HPI. Contents Basic Catalogue [EB/OL]. 2000. [2019-05-01]. http: //www. jtekt-hpi. com/wp-content/uploads/2015/09/CATALOGUE-TECHNIQUE-HPI. pdf.

[71] AXOR. Microspeed data sheet [EB/OL]. [2019-05-01]. http: //www. motioncontrolproducts. com/pdfs/MCS-plus-servo-drive. pdf.

[72] Sunhydraulics . DDDC four-way valve data sheet [EB/OL]. [2019-05-01]. https: //www. sunhydraulics. com/model/DDDC.

[73] Sunhydraulics . RBAE relief valve data sheet [EB/OL]. [2019-05-01]. https: //www. sunhydraulics. com/model/RBAE.

[74] Sunhydraulics. NCCC needle valve data sheet [EB/OL]. [2019-05-01]. https: //www. sunhydraulics. com/model/NCCC.

[75] APH. C10 18 series cylinder data sheet [EB/OL]. [2019-05-01]. http: //www. aph. co. uk/

wp-content/uploads/2017/04/APH-C10-Brochure. pdf.

[76] Sakae. CFL200 potentiometer data sheet [EB/OL]. [2019-05-01]. http: //www. meditronik. com. pl/doc/bourns/syp087103. pdf.

[77] Omega. PXM309 pressure transducer data sheet [EB/OL]. [2019-05-01]. http: //www. omega. com/pressure/pdf/PXM309. pdf.

[78] Quanser. Q8-USB data acquisition device data sheet [EB/OL]. [2019-05-01]. https: // www. quanser. com/wp-content/uploads/2017/04/Quanser_ Peripherals_ high_ res_ Final. pdf.

[79] RANA A S, SAYLES R S. An experimental study on the friction behaviour of aircraft hydraulic actuator elastomeric reciprocating seals [J]. Tribology and Interface Engineering Series, 2005, 48 (5): 507-515.

[80] BONCHIS A, CORKE P I, RYE D C. A pressure-based, velocity independent, friction model for asymmetric hydraulic cylinders [C] //Proceedings 1999 IEEE International Conference on Robotics and Automation(Cat. No. 99ch36288c).United States: IEEE, 1999, 3: 1746-1751.

[81] TUNAY I, KAYNAK O. Provident control of an electro-hydraulic servo system [C] // Proceedings of IECON' 93-19th Annual Conference of IEEE Industrial Electronics. United States: IEEE, 1993: 91-96.

[82] SCOTT K. Handbook of Industrial Membranes [M]. NL: Elsevier Science, 1998: 271-305.

[83] YANADA H, TAKAHASHI K, MATSUI A. Identification of dynamic parameters of modified LuGre model and application to hydraulic actuator [J]. Transactions of the Japan Fluid Power System Society, 2009, 40 (4): 57-64.

[84] TRAN X B, HAFIZAH N, YANADA H. Modeling of dynamic friction behaviors of hydraulic cylinders [J]. Mechatronics, 2012, 22 (1): 65-75.

[85] FITCH E C, HONG I T. Hydraulic component design and selection [M]. Stillwater, OK, USA: BarDyne, 1998.

[86] DU C, PLUMMER A R, JOHNSTON D N. Performance analysis of a new energy-efficient variable supply pressure electro-hydraulic motion control method [J]. Control Engineering Practice, 2017, 60: 87-98.